A DOE Handbook:

A Simple Approach to Basic Statistical Design of Experiments

Daniel Coleman and Bert Gunter

May 14, 2014

Contents

Foreword

Why Yet Another Statistical Design of Experiments (DOE) Book?

The simple answer is, of course, that we think there's still a need. Basic multifactor DOE is still not as widely practiced as it should be, especially in industry and technology. There are many reasons for this, but we believe that a large part of it is that the standard methods and strategies are too complex and arcane for many practitioners to readily digest and apply. We consider this a major loss.

And it need not be. In particular, we think that some (surprising) discoveries in the 1990's offer a real opportunity to change this reality; this handbook is an expression of this conviction. These discoveries concern the "projectability" of 2-level orthogonal designs and are discussed in detail in Chapter 3. They provide the basis for a simple framework for the design and analysis of 2-level, multifactor designs that we hope encourages wider application in science and technology, especially in industry. So far as we know, no other book takes this approach.

We deliberately refer to this short text as a "handbook" to emphasize its simplicity. Most DOE texts extend to hundreds of pages. Of course, this comparison is unfair, because we have not attempted to cover the range of topics that they do. There are several reasons for this. First and foremost, we believe that simplicity is a virtue, and that effective application of simple statistical methods by subject matter experts with deep insight into the systems and processes they study yields greater gains than complex methods in the hands of statistical specialists. So we explicitly limited our goal here to bringing to experimenters the basic DOE methods that we believe offer major improvements in experimental efficiency and information, avoiding any technical niceties that might dilute this effort.

Second, as we have already emphasized, the projectability-based approach is inherently simple. This enables us to avoid a lot of mathematics that existing texts must spend time on.

Finally, we frankly think that the emphasis on statistical inference that occupies many texts distracts from the central scientific learning task. As many have remarked, the power of DOE is in the *design*, not the analysis. A well-designed experiment is typically straightforward to analyze; while often one that is poorly designed (or executed) will confound any effort to make sense of the results. We have covered what we consider to be the essentials of design and execution that lead to informative, interpretable results. While there are certainly circumstances where more than these bare essentials are necessary – and we have pointed some of them out in the text – our priority has been to maintain simplicity and accessibility. This is another reason that we have called this a *handbook*, not a textbook.

Our views on these issues have been largely shaped by our experience in industry over several decades of practice, both as collaborators in the design and analysis of industrial experiments and as teachers of DOE to scientists and engineers in a variety of industries and disciplines. We have learned a lot from this, not only about many fascinating areas of science and technology, but also about the practical realities of industrial experimentation that challenge both imagination and technical acumen. We have come to appreciate the difficulty of planning and executing successful experiments in environments that are never quite as controlled as one would like; within time frames that are always too short; and with limited resources that have to be juggled to fit. We hope that this handbook contributes to the efforts of experimenters to better understand and improve the processes and products on which they work and that constitute the technology, goods, and services that make up the fabric of all our lives.

Acknowledgement and Disclaimer

This handbook grew out of our work as employees of Genentech, a major biotechnology company that is a member of the Roche Group. Responding to requests for help in experimental design, our (now former) manager, David Giltinan, asked that we develop and deliver basic DOE training to the company's engineers and scientists. We did this, but discovered that when participants in our training requested an accessible basic book to use as a resource and reference to accompany our class notes, we could find nothing that we and they agreed was suitable. So we decided to write one. We would therefore like to thank both David and our students for encouraging our efforts – this would not have happened without them.

Naturally, many of the examples in this handbook have been informed by our work at Genentech. However, we need to explicitly note that except for the one example taken from the literature, all the other examples are either outright fabrications or have been modified to more clearly demonstrate the points we wished to highlight. In particular, the experiments we describe and the results "obtained" may strike subject matter experts familiar with the issues as unusual or incorrect. That is entirely due to us as authors. Had we related accurate, complete accounts of any real experiments that motivated our examples, they would have reflected best practice. We are also of course obligated not to reveal any proprietary information, and circumstances and data have been altered for that reason. Nevertheless, we hope that we have provided sufficient versimilitude to show how DOE applies to real problems even within these constraints.

Chapter 1

Introduction

1.1 What is Statistical Design of Experiments –DOE – And Why is it Useful?

One of the hallmarks of the "scientific method" is that all theory must be validated by data: Mother Nature must always be the ultimate arbiter, a concept that was one of the great breakthroughs of Western philosophy in the Renaissance. It clearly separated theology, art, and even mathematics, which required only internal coherence and beauty to be valid, from the essential empiricism of science.

Scientific data are obtained in two basically different ways: through passive observation and active experimentation. Examples of the former are astronomical observations, weather measurements, geographical patterns of disease incidence, and earthquake seismometer readings. Obviously, such data are essential for determining how Nature works and serve both as raw material from which new models – that is, new theories – are built and as evidence for or against existing ones.

It is the active experiment, however, that was undoubtedly the greatest development of the scientific revolution and the catalyst for the enormous role that science and technology play in shaping human activity. Instead of passively awaiting informative events, Mother Nature is deliberately prodded to reveal Her secrets. After all, Galileo didn't merely assert that light and heavy objects fell at the same rate; he actually dropped them from the Leaning Tower and compared their times of fall (or so legend has it). What makes active experimentation so powerful is that it can establish *causality*, by which we informally mean a physical relationship between observed phenomena and active agents (their "causes").[1] Passive observation can do this, too – for example, many of Einstein's predictions of General Relativity were verified by astronomical observations – but the ability to actively control the observational environment through experiment is a hallmark of modern science and engineering.

Broadly speaking, there are three qualitatively different kinds of experimentation. The first, and the one that one reads about in textbooks and makes the headlines, is what one might call the conceptual breakthrough. Examples are Lavoisier's discovery that water was composed of hydrogen and oxygen, Jenner's cowpox vaccine for smallpox, or Bednorz and Müller's discovery of high temperature superconductivity. These were all clear, definitive results that completely on their own established or contradicted a scientific paradigm. However, such "game-changing" discoveries are relatively rare in science and engineering practice, and they are *not* the sort of experimentation that will be discussed in this book.

Virtual experiments through computer simulation is a second type of experiment that is a relatively recent addition to the scientific lexicon. In some respects, this is not experimentation at all, because it involves no actual physical manipulation of Nature. Instead, physical (including biological, economic, etc.) phenomena are modeled mathematically and parameters of the mathematical model are changed to determine the effect on the computed results. While such simulations do, in fact, exhibit many of the characteristics of real physical experimentation, the specialized statistical methods required to deal with them are again not our concern in this handbook.

The experimental context that *is* our concern here is perhaps best described as physical experimentation whose purpose is to improve exist-

---

[1]A more careful discussion of this is certainly required, but would take us too far afield into the philosophy of science.

ing processes and incrementally extend current knowledge. While perhaps somewhat less glamorous than the previous sorts of experiments, such efforts comprise the bulk of scientific and engineering work, especially in industry. Examples include ways to improve the sensitivity of a measurement by manipulating ("optimizing") the conditions and settings used; raising the temperature for superconductivity of a high temperature superconductor by modifying the composition of the material; or improving the yield of a biochemical process by altering process conditions. The kind of evolutionary improvement that result from such experimentation brings us faster computers, better drugs, sweeter tomatoes, and more efficient cars. In short, they are woven into the fabric of our lives.

The essential feature of such experiments is the need to manipulate and evaluate the effects of possibly many different controllable inputs, commonly referred to as experimental factors or parameters (we prefer "factor" because "parameter" means something else in statistics). What makes this challenging is that experimentation can be expensive and time consuming, and experimental "noise" can make it difficult to clearly understand and interpret what results. These are the practical difficulties that DOE was developed to address.

1.2 Why is DOE Useful?

Perhaps the most concise and insightful explanation of the value of statistical design of experiments was given by one of its foremost developers and proponents, George E. P. Box, who said that it "catalyzed the scientific learning process." What this means is that the use of DOE methods helps experimenters obtain better information faster and with less experimental effort. In the real world where experimental resources are limited, this can often be the key to making progress.

The essential ideas that justify such claims are simple and compelling. In fact, there are really only two principles at work: the first is that one gains more precise information on the effect of a deliberate change by making the change many times and basing conclusions on the average[2] of the results, rather than by making the change just once and using just that single result. The second is that in order to evaluate possible joint, interactive effects of multiple experimental factors one must actually simultaneously change them and see whether and how they interact.

While these statements may seem obvious, it is nevertheless the case that standard experimental practice often violates them. To see how, consider the following simple artificial example.

1.2.1   OFAT Versus Factorial Designs: Main Effects

An experimenter wishes to investigate how to improve the oil yield in an algae growing process for biofuels by adjusting the nitrogen feed rate and temperature. To do this, she decides to produce a batch of algae at each of five separate operating conditions that cover what she considers to be a reasonable range of possibilities and measure the lipid concentration (in gms/liter) that results for each. The five conditions are:

1. Control. Use current standard feed rates and temperatures.

2. Nitrogen +. Increase the nitrogen feed rate 5% from control.

3. Nitrogen −. Decrease the feed rate 5% from control.

4. Temperature +. Increase the temperature 1.5° from control.

5. Temperature −. Decrease the temperature 1.5° from control.

If "0" is used to label control and "−" and "+" are respectively used for the "decreased" and "increased" settings, this set of conditions – the experimental "design" – can be conveniently represented as the following table, where each row gives the settings for a single experimental "run", or condition:

---

[2]or other appropriate summary

Table 1.1: Algae Experimental Design 1

Condition	Feed Rate	Temperature	Yield
1	0	0	
2	+	0	
3	−	0	
4	0	+	
5	0	−	

Of course, the "Yield" column would be filled in with the results once they were obtained. This kind of design is known as "one factor at a time" – or "OFAT" for short – because the changes from control are made in only one factor at a time in order to clearly evaluate their effect. For example, suppose the results were:

Table 1.2: Algae Experimental Design 1: Results

Condition	Feed Rate	Temperature	Yield(gms/liter)
1	0	0	22
2	+	0	23
3	−	0	18
4	0	+	20
5	0	−	22

One might reasonably conclude based on the large yield increase from 18 to 23 when the feed rate changes from 5% below to 5% above control that the higher feed rate is desirable. On the other hand, changing temperature appears to have little effect, with perhaps a slight yield decrease when temperature is increased, although this might just be due to experimental variability, such as variations in the equipment operation, raw materials, condition of the algae seed stock used in each condition, and so forth. There might also be some uncertainty in the measurement of yield that contributes to the variation in these values. In any case, based on these results it seems reasonable to increase the feed rate but leave the temperature alone, which is the condition of run 2. Note that these conclusions are each based on a single pair of runs: 2 vs 3 for feed rate and 4 vs 5 for temperature. No replication or averaging of any sort has been done, thus violating the first of the cardinal design principles; hence confirming the conclusions by re-running condition 2 would obviously be desirable.

More subtly, by changing only one factor at a time, there is no possibility of determining whether temperature and feed rate might interact in some way. If possible, we would prefer to have some way to test for such possibilities, hopefully without a large amount of additional experimentation. It turns out that DOE provides a way to do this *and* simultaneously assess results by averaging multiple runs, rather than relying on unique results. Moreover, it can be done with little or no additional experimental effort. To show how, it is helpful to recast the experimental design in geometric terms as a pattern of points in the plane.

To do this, merely convert the symbols "+" and "−" to +1 and −1. We can now plot the design settings of Table 1.1 in Figure 1.1.

Figure 1.1: Initial Design to Study Feed Rate And Temperature Effects on Algal Growth

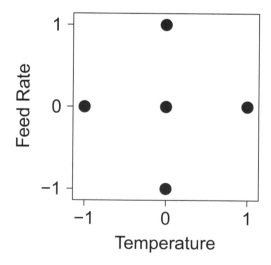

It turns out that a better design that enables us to accompish both our objectives is to move the four points at the centers of each edge to the corners. as shown in Figure 1.2. This improved design is known as a 2×2, 2^2, or full factorial design with center point.

Figure 1.2: Improved Biofuel Yield Improvement Experimental Design

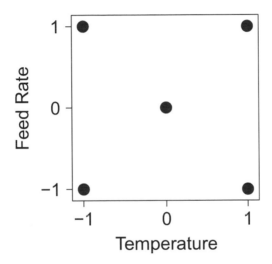

Now suppose that this design were run. Using the same underlying process model that produced the OFAT data[3], the following were the results:

---

[3]$yield = 21 + 2.5 \times FR - 1.5 \times Temp + 2.0 \times FR \times Temp + error$, in the (-1,1) scaling, rounded to whole numbers

Table 1.3: Algae Experimental Design 2: Results

Condition	Feed Rate	Temperature	Yield
1	0	0	22
2	−	−	22
3	−	+	14
4	+	−	20
5	+	+	25

To determine the effect of changing the feed rate, one can see from the plot or the table that there are now two differences to consider: first, on the left in the plot when temperature is low – conditions 2 and 4 in the table; and then again on the right in the plots when the temperature is high – conditions 3 and 5 in the table. At the low temperature, the feed rate increase results in a yield decrease of 2 gms/liter (22 to 20). At the high temperature, the yield increases by 11 (from 14 to 25). We thus have two separate sub-experiments whose results can be averaged to give an average yield increase of (-2 + 11)/2 = 4.5 g/l. Note that this average effect is based on twice the information – two separate changes instead of just one – as that of the OFAT.

In an exactly analogous way, there are also two sub-experiments comparing yield results when the temperature is changed: the bottom pair of points in the design plot when feed rate is low (conditions 2 and 3 in the table) and the top pair when feed rate is high (conditions 4 and 5). With low feed rate, the temperature increase is associated with a decrease of 8 g/l; with high feed rate, the yield increases by 5 g/l. Again, these two separate results can be combined to give an average yield *decrease* of (-8 + 5)/2 = -1.5 g/l . As before, this result is based on twice as much information of the OFAT experiment. These results thus make a stronger case that feed rate should be increased and now also that temperature should be decreased to improve yield.

But there is something more going on here. From the table, it appears that, in fact, higher yields are obtained when both temperature and feed rate are increased to their "+" setttings (20 versus 25 g/l). Thus far, we have only calculated the *individual average* effects of the two experimental factors, called the *main effects* of the two factors. However, experimenting at the corners of the square rather than on the sides also allows assessment of joint *interactive* effects. As we shall see in a moment, this turns out to be important here.

While these effect calculations may seem tricky, there is nothing illegitimate about them. The results at all four corners of the design square have been reused for both feedrate and temperature main effect assessments. For the OFAT experiment, only the two points on the sides were used to assess the effect of changing the factor settings. By using a design that forces the data to do such multiple duty, more information is obtained with the same experimental effort. This is one of the principal strengths of DOE.

Finally, we note that these calculations did not make any use of the center/control point value of 22. In fact, neither did the OFAT design, as the information on the effects was based essentially on the yield difference between the low and high settings on the sides. Since the center point apparently contributes no useful information, it seems as if it could be omitted. However, we shall argue in Chapters 2 and 3 that not only is running the center point a good idea, but that repeating it more than once is usually worthwhile. We shall expand on these matters there.

1.2.2 OFAT Versus Factorial Designs: Interactions

The OFAT design gives us no information on interactions, because feed rate and temperature are never simultaneously changed from their control settings. But the factorial design does make such changes, and this gives it the ability to probe these joint effects *above and beyond* what would be expected from their main effects.

The bioprocess experiment was constructed to clearly show this. There are two equivalent ways of looking at the interactive effect of feed rate and temperature. First, consider the temperature effect. We saw above that there are actually two mini-experiments in which temperature changed, when feed rate was low and when feed rate was high, resulting in a decrease of 8 g/l in the first case and an increase of 5 g/l in the second. To calculate the main effect, we averaged these two results, but we can also ask: Is there a difference in the effect of changing temperature depending on what the feed rate is? This is important, because, it would say that you can't really talk about the effect of temperature in isolation – it might depend on what the feed rate is. Or, in other words, their effect on yield needs to be described jointly. To answer this question, we clearly need to consider the *difference* in the temperature effect between the two feed rates, which is 5 g/l – (-8 g/l) = 13 g/l (where "-" means "decrease"). For technical reasons – to normalize all effects to the same scale – we want this, like the average, to be divided by 2, so we see that the temperature×feed rate interaction is 6.5 g/l, even larger in magnitude than the feed rate main effect of 4.5 g/l that we previously calculated. So there does seem to be a rather large interactive effect: the effect of changing temperature does appear to depend on what the feed rate is.

One can equivalently approach this by asking the question as: (how) does the effect of feed rate depend on temperature? When temperature is low, we saw previously that the effect of changing feed rate from low to high was a 2 g/l decrease in yield; when the temperature is high there was an increase 11 g/l. So the difference in the effects from low to high temperatures is 11– (-2) = 13 g/l, which we halve as before. Mathematically, the two ways of describing the interaction are the same: it can be expressed whichever way seems more meaningful or convenient. But the point is that one cannot talk about the effect of changing either factor in isolation. The effect of changing one depends on what the value of the other is.

When the effect of the interaction is taken into account, it clearly changes what should be done to maximize yield: the feed rate should be increased, but temperature should be increased, not decreased, to take advantage of the large increase in yield when temperature is high at the high feed rate versus when it is low. So the "+" "+" setting in both factors is where we expect maximum yield to occur. Note also, that somewhat better results would be expected here than if temperature were kept at the "0" setting that seemed optimal based on the OFAT design.

Of course, this example was constructed in order to show the importance of learning about interactions. In many systems, interactions among the factors (over the ranges in which the factors were changed) are relatively small and can be neglected. But, you can't know this without explicitly testing for it, which OFAT experiments do not. OFAT experiments implicitly *assume* interactions are unimportant, while the designs given in this book allow the data to tell their own story. In addition to the improved precision due to averaging, This is another virtue of DOE.

1.3 Organization of this book

DOE is a well developed discipline with an extensive literature. R.A. Fisher developed the essential ideas in the 1920's, and his ground-breaking book, *The Design of Experiments* (Fisher, 1935), was originally published in 1935; but DOE remains an active area of research and publication today. Our goal here is to present only a small subset that our experience has shown us to be a useful "toolkit" for a broad spectrum of experimenters. While mostly this has tended to mean engineers and scientists working in industrial applications, DOE has also been effectively used in measurement process development (e.g. of medical and environmental assays), business process improvement (e.g. reducing transaction times), and large scale computer modeling and simulation (e.g. of ecological systems), among other areas.

To keep the exposition as accessible as possible, we avoid mathematical formulations and rigorous justification of why and how the methods work, leaving such details to the references for those who are curious. Fortunately, such complexity is not usually necessary for effective use. Instead, we describe the underlying concepts non-mathematically, provide a catalogue of designs and instructions for selecting those that best fit the situation at hand (including when to seek help when none may suit), and rely almost exclusively on graphics for analysis and presentation of results.

Chapter 2 provides what we see as the essential underpinnings for *any* good experiment. Here, we discuss the underlying structure and terminology of an experiment, especially the notion of the *experimental unit,* a topic rarely discussed in DOE textbooks but one that we have found frequently confounds experimenters and compromises experimental results. We shall also discuss *randomization* and *blinding* in experimentation and measurement to avoid spurious biases, another topic which we feel does not receive adequate attention in standard expositions.

Often complete randomization is not a good idea or may even be logistically impossible. In these situations, *blocking* and *split-plotting* can be employed. Unfortunately, these are complex topics for which even a modest exposition would take us too far afield. So our goal here is just to make experimenters aware of the concepts and the challenges they present in experimental execution and interpretation. This is one of the situations where the best advice may be: if you think these issues matter, seek expert statistical help.

We also take up the idea of experimental replication, and especially the difference between "true" replicates and what we (and others) refer to as "bang-bang" or "technical" replicates. Confusion between these two kinds of experimental repeats can lead to irreproducible experimental results or the inability to translate pilot scale results to full scale industrial processes..

In Chapter 3 we describe the central concept on which we base our simple approach to DOE: design "projectability." Relatively recent developments in this area (*ca.* 1995 and later) allow us to provide a simple, but useful catalog of designs. While we certainly do not claim that this approach is inherently better than others that are widely used, we have found that it works well and is straightforward to learn and use.

Finally, in Chapter 4 we show how to analyze and interpret experimental results. One difficulty here is that software is typically used for the analyses, and there are many DOE software packages and add-ons available. So we shall keep the presentation as generic as possible, relying on informative graphics and simple calculations that even standard spreadsheet software can handle.

1.4 A word about "p-values"

Readers with some prior statistical exposure may be surprised that we say nothing about "statistical significance" or "p-values" in this handbook. These are statistical concepts that are widely used to confer a sort of statistical legitimacy on data analyses. However, there is considerable controversy about whether this practice is really such a good idea. Our view is that for the kind of *exploratory* experiments that we are concerned with in this book – as opposed to the *confirmatory* experiments of, say, medical clinical trials – statistical significance and its associated principles aren't relevant and may even obfuscate, because their use presupposes a formalism that in most cases doesn't exist. So we do not use them. Those who feel deprived by their omission should search elsewhere for comfort.

1.5 A word about references

Due to the ubiquity of Internet search engines, Wikipedia™, online reviews, and freely accessible online courses, tutorials, and even research publications, we have decided to restrict our reference list to short bibliographies at the end of each chapter. These will give citations for specific references in the text, relevant research publications that fill in essential technical background, and some personal favorite DOE resources of ours for books and websites. But we encourage readers to search on their own, as there is sure to be more out there to fill in background, give examples, and even provide software for implementation.

1.6 References and Resources

Fisher, R.A. (1935). *The Design of Experiments*, Oliver and Boyd.

Works of George Box

George Box (1919 - 2013) was one of the major developers of DOE and was especially concerned with its effective application in science and technology. He and his collaborators wrote several classic textbooks that we think are still very good general practical DOE resources. However, even if you are not interested in a comprehensive technical exposition, Box and his co-authors devoted space in their books to thought provoking discussions of the role of DOE in scientific and industrial development and practical considerations and challenges in implementation. In our view, these discussions are worth reading even if you skip everything else. Here are the books and relevant sections.

Box, G.E.P., Hunter, J.S., and Hunter, W.G. (2005). *Statistics for Experimenters: Design, Innovation, and Discovery*, 2nd ed.,Wiley, Hoboken.

Especially: Chapters 1; Section 5.12 on "Dealing with more than one response"; The introduction and first section of Chapter 11 on "Modeling, Geometry, and Experimental Design."

Box, G.E.P. and Draper, N.R. (1987), *Empirical Model Building and Response Surfaces*, Wiley, New York.

Especially: Chapter 1 and Chapter 14, which discusses why statistical concepts of "optimal" experimental designs may not be practically useful.

Box, G.E.P. (2006). *Improving Almost Everything: Ideas and Essays, Revised Edition*, Wiley, Hoboken.

This is a collection of mostly nontechnical essays discussing the application of statistics generally, and experimental design particularly, to industrial quality and process improvement.

Websites

The best way to find useful DOE websites is to search for them ("Design of Experiments," "Statistical Experimental Design," and variants thereon are keywords that bring up lots of relevant hits). However, one particularly useful site worth highlighting is the NIST/SEMATECH Engineering Statistics Handbook at http://www.itl.nist.gov/div898/handbook. Module 5 ("Improve") is the one on DOE, but there is much of interest in other modules, also. Be forewarned, however: the handbook is classical in its approach and may be somewhat dated in places. Nevertheless, we still think it is a useful resource worth perusing.

Chapter 2

Effective Experimentation

The goal of any experiment is to obtain clear, reproducible results. By this, we mean that the questions the experiment was designed to answer are answered; and that conclusions from the experiment should hold not just for the particular lab, time, equipment, raw materials, reagents, methodology, and personnel that comprised the experimental "environment" in which the experiment was performed, but for *any* lab following "similar" methods, using "similar" equipment, raw materials, reagents, personnel, and so forth. Of course, exactly what is meant by "similar" can be tricky to specify. After all, no two pieces of laboratory or measuring equipment are exactly the same; and even a single device changes over time; raw materials and "pure" reagents vary somewhat from lot to lot; measurement equipment is never calibrated in exactly the same way; people vary in training and background. As every experienced experimenter knows, sometimes what appear to be subtle, minor differences can have an unsuspected large effect on experimental results. But the goal remains to establish general scientific validity; it isn't science if the results only occur once or work only for one experimenter.

Beyond this, in many engineering and industrial experiments the goal is to generalize what is learned from small, laboratory-scale experiments to full scale engineering applications or industrial processes. This adds additional complications, especially in translating the essential features of the changes made in the lab to their full scale equivalents. For example, suppose one performs an experiment to determine optimal temperature, air flow, and cure times for curing an anti-glare coating on glass. The goal is to get a full cure as rapidly as possible without introducing defects such as pinholes or bubbles in the coating. How does one translate the optimal temperatures, flows, and times found from experimenting in a few cubic liter laboratory oven to the full scale production operation which occurs in a continuously fed, open annealing oven that is thousands of times larger in volume with very different heat transfer and airflow properties? To do this successfully requires subject matter expertise in order to understand the relationship of the experimental settings at the lab scale to the fundamental heat transfer properties of interest so that these can be replicated at full scale. Needless to say, this can be difficult.

In addition to these inherent challenges, if experimental results are tainted by systematic biases, high experimental variability, or outright errors due to poor design or bad procedures, no statistical analysis can recover the missing information to produce reliable conclusions. Worse yet, if the problems aren't recognized, wrong conclusions might result. It is therefore important to understand how this might happen, not only to avoid it in one's own experiments, but also to help evaluate the validity of others.

We therefore want to discuss some key *operational* characteristics that we think a good experiment should have. We call this topic *effective experimentation*. Among these, we would certainly include subject matter expertise, "GLP" (Good Laboratory Practices), proper equipment maintenance and calibration, and so forth, none of which have anything to do with statistics. These are the hopefully routine but vital practices required to maintain the experimental environment so that consistent reproducible results can be (safely) obtained.

But there are other operational aspects of experimental practice that do intersect statistical design and analysis and which, if not followed, can also lead to inconsistent, irreproducible results. Among these are randomization, blinding, proper replication, blocking, and split plotting. These are the topics of this chapter.

The common thread running through them – and what makes them "statistical" in nature – is that they are concerned with how appropriate conclusions can be drawn in the presence of different kinds of experimental "error". What is meant by this is **not** what one might call experimental blunders, such as a wrong amount of reagent added to a test tube, but the uncontrollable influences of environment, equipment, materials, etc. that make it impossible to perfectly replicate results when experiments are repeated. Mistakes can be fixed,[1] but these sorts of experimental errors cannot be, at least not without making major changes to the experimental setup. They are better thought of as inherent variability in the overall experimental environment, but the "error" terminology is standard, and we shall use it.[2] Examples include variation in individual pieces of equipment and reagent lots, slight variability in the measured amounts of reactants, positional effects in ovens, differences among expert raters of images, and so forth.

As we shall see, there are techniques that can be incorporated into the way an experiment is conducted that can prevent, or at least reduce the risk that these various sources of variability corrupt experimental results. That is the theme of this chapter.

2.1   Some terminology

To begin with, we need to define some terminology. We especially need to avoid miscommunication: different disciplines use different terms to describe similar experimental concepts, and this can easily lead to confusion. We shall follow the terminology we define here and leave it to readers to translate into the "native" language of their choice.

As we have emphasized already, an experiment is a process in which deliberate changes are made in experimental *variables* or *factors* (these mean the same thing) to investigate their effect on one or more measured *responses*, the outcomes of interest.

---

[1]In quality control, these are often referred to as "assignable causes."

[2]In quality control, these are often referred to as "common causes.

The intent here is to establish *causality,* in order to assert – and usually quantify – a physical link between the controlled change in the experimental factor and the observed change in the experimental response. As noted in the Introduction, this deliberate change is the defining characteristic that distinguishes active experimentation from passive observation.

For example, in the biofuels example of the Introduction, the experimental factors were temperature and nitrogen feed rate, and the response was the lipid yield in gms/liter. In an experiment to improve the tensile strength of a polymer fiber, the experimental factors might be which of two different organic solutions the polymer fibers are drawn through, the temperature of the solution, the rate of draw, and the type of coating on the fiber. In an experiment to examine the toxicity of a chemical compound on fish, experimental factors might be concentration of the compound, pH of the water, species of fish, age of the fish (e.g. hatchlings or adults), water temperature, and length of exposure. The response might be proportion of fish that die or exhibit skin lesions.

The values at which the experimental factors are controlled are called the *levels* of the factors. Note that such levels can either be discrete, categorical entities, such as type of organic solution in the polymer experiment or species of fish in the toxicity study; or a particular value chosen from a continuous range of possibilities like solution temperature or water pH. Sometimes one even has a mix of these two possibilities, like "hardness" on a discrete scale of 1 to 5 or size among "small," "medium," and "large." These possibilities are called "ordered categorical factors" or simply "ordinal factors" to emphasize that the different possible settings of the factor have a natural ordering, but it's not really possible to control them on a fine scale. Note how this differs from a factor like fish species, which simply has different labels without any meaningful ordering.

Responses also can come in these different variable types. Generally, one prefers to have more informative continuous measured responses like lipid yield, tensile strength, or mortality rate = proportion that die. But note that if there were only 5 fish in each tested group, then the mortality rate is essentially an ordinal variable with 6 possible values (0 to 6). One might even have just a 2 value categorical response like good or bad, or colored or not, e.g. when experimenting to determine what produces off-color in what should be a transparent product. Not surprisingly, the more informative the response, the less experimentation one generally needs to do to understand what is going on.

When experimenting with many factors, each experimental condition must specify the settings of all factors. Each such specification is called an experimental *run*; the collection of all runs that are to be made constitutes the *design* of the overall experiment. For example, if there are 5 factors, then each run requires specifying the levels for all 5. If there are 12 runs in the overall design, there are 12 separate sets of five levels.

We summarize all this nomenclature and give examples of each from the biofuels experiment by:

Response Measured experimental outcome (lipid yield in g/l)

Factor An experimental variable that is changed in a controlled manner (temperature, N feed rate)

Level Values at which a factor is controlled (Temp: Ctrl, Ctrl $\pm$ 1.5°; N feed rate: Ctrl, Ctrl $\pm$ 5%)

Run The combined setting for all the factors at which a response is measured (Temp = Ctrl + 1.5°; feed rate = Ctrl − 5%)

Design Set of all experimental runs (2^2Factorial + Center − See Introduction)

Finally, we mention one more somewhat esoteric bit of terminology that can be a source of confusion. Formal statistical analysis of experimental results typically involves the use of a statistical technique known as "linear modeling" *aka* "multiple linear regression". Results from such an analysis are expressed in the form of simple equations that *fit* the experimental data and try to *predict* what the the measured responses will be based on the experimental factor settings (and hence are referred to as *fitted models* or *prediction equations*), e.g.

$Lipid = 20.6 + 2.0 \times Temp + 2.5 \times FeedRate$

It's important to to specify the scaling here: Lipid is measured in g/l and both Temperature and Nitrogen feed rate are expressed on a coded scale where -1 means the low level of either factor, 0 means the middle control level, and $+1$ means the high level. It is easy to convert any actual value of temperature or feed rate to this coded scale and vice-versa. For example, a temperature of Ctrl $+1°$ corresponds to $+2/3$ on the coded scale, and $-.5$ on the coded feed rate scale corresponds to a feed rate of Ctrl -2.5%. It is also easy to re-express the coefficients of 20.6, 2.0 and 2.5 to equivalent values if Temp and FeedRate are measured on their actual scales. But these are unimportant details that we don't wish to bother with here, so we just use the coded scale now for convenience.

In the regression fitting procedure used to derive an equation such as this, Temperature and FeedRate values are the known experimental values at each of the (in the example, 5) runs of the design, and the coefficients are the *unknown "parameters"* that the procedure "fits" (or "estimates"*). And therein lies the source of possible confusion. In many science and engineering disciplines, the term *parameters* is used to refer to the Temperature and FeedRate experimental *factors*, rather than to their fitted coefficients. We shall adhere to the statistical terminology: in this handbook, *factors* refers to the experimental variables; *parameters* refers to the coefficients of the prediction equation that are fit from the experimental results.

2.2   The Experimental Unit

Example 1: A Xenograft Study

An animal experiment was being planned to evaluate the effect of a proposed new drug on a type of cancer. The protocol for the study was to implant a human tumor – a so-called xenograft – in each experimental animal, rats in this case, and then measure tumor growth (or lack of it) in the drug treated versus untreated animals. Part of the discussion was how many rats would be needed to make a convincing case that the drug worked, if indeed it did. As might be expected, individual animals vary both in their response to the tumor (e.g. due

to differences in their immune systems) and drug (for rats in the drug
treated group). So such a discussion turns out to be about the ability
to detect a minimal meaningful difference in the tumor growth rate
between the two treatment groups. The details of what is meant by
"meaningful" and "detect" are, of course, important, but sorting this
out carefully would involve complex statistical issues that we wish to
avoid. For the purposes of the discussion here, we merely state what
should make some intuitive sense: the larger the sample size of tumor
bearing rats, the greater is this ability. A crucial sticking point that
arose, however, turned out to be exactly how one counted tumors to
determine what was meant by "sample size."

The contract animal laboratory running the study said that it would
be possible to implant two tumors in each rat and thereby double
the sample size and reduce the number of animals required to achieve
the desired detection level. Some people involved in the discussion
greeted this idea with enthusiasm, as it would result in considerable
cost reduction (animal testing is expensive!). Others weren't so sure,
however, as it sounded like they would be getting something for almost
nothing, always a cause for suspicion and even skepticism.

Example 2: A Fermentation Study

Bioprocess engineers wanted to compare two different types of a reagent
used in a fermentation process to determine which produced a higher
yield of the purified protein that was the end product. Two batches of
the fermentation culture were prepared, one each with each reagent,
and then each batch was split among 8 different 2 liter "mini"- fermenters in which the experimental fermentations were done. The protein of interest was then purified from each fermenter to determine the
yield, giving 8 yield results for each type of reagent. As usual, even
within each group of 8, the yields varied somewhat from mini-ferm to
mini-ferm, so a standard statistical procedure was used to determine if
the average yield difference between the groups of 8 was large enough
to be statistically "significant."

... Until, that is, a statistician told them that no useful conclusions
at all could be drawn from the results. This announcement was not

well received, as the overall experiment had already taken well over a month to perform.

What's Going On?

It turns out that the key issue common to both these examples is the concept of – and confusion about – the experimental unit, for which a definition is:

> **Experimental Unit** The smallest division of experimental material that can receive separate treatments.

But what does this mean? – and why is it central in both these examples?

Let's first consider the xenograft experiment. The experimental "material" here are the animals, and the experimental unit is therefore the individual rat: individual rats receive different treatments, but a single rat is either treated with drug or not. "Sample size" always refers to the number of experimental units, so the two tumors per animal should actually be considered duplicate *measurements* of tumor growth rate per individual experimental unit. There is some value to duplicating the measurements and using their average as the individual per rat result; but it does not double the sample size and (usually) does not double the power to detect experimental factor effects.

This also makes biological sense. After all, both the cancer and drug affect each rat as a whole. If a rat has a bit stronger immune system, both tumors will grow more slowly; if its liver tends to break down the drug more quickly, the drug will be less effective and both tumors will grow more rapidly. In other words, the tumor growth of the two tumors within each animal are likely to be more similar than tumor growth between different animals. This means that the growth of the two tumors within an animal really aren't separate – or *independent* – assessments: it's more like the same growth twice. This is not double the information!

What is the experimental unit in the second example? The treatment here is reagent type, and the experimental unit is the prepared culture batch in which the reagent is used. Every time a different batch is prepared, a choice of reagent can be made. But once this is done, no further change can be made to the sub-batches split up among the fermenters. This means that there were actually only two experimental units here, one with one reagent and the other with the second. One can think of the multi-week process of fermenting the batch in 8 different fermenters and then purifying and measuring the amount of protein that resulted as one giant measurement process. So what we really have here is 8 measurements of each reagent batch!

And that's the problem that the statistician complained about: the study can give no information about how much variation there might be from one batch preparation to another *within* reagent types. Without knowing this, it is impossible to know whether any difference seen between the two groups of fermenters is due to the different reagents or just to batch preparation to batch preparation differences that might be seen with a single reagent. One can hope it is the former and not the latter, but that can be a mighty leaky boat in which to float an expensive decision.

These examples hint at many subtleties that might be involved in deciding how to trade off the costs and benefits of more experimental units versus taking more "measurements" per unit. For example, if batch preparation were a relatively simple and cheap matter, then preparing 16 separate batches, 8 with each reagent type, and fermenting one batch in each fermenter might be the way to proceed. But if the availability of some other component of the batch were severely limited, this might not even be possible! One therefore might imagine some kind of hybrid strategy in which, say, 2 batches of each reagent type were prepared and then split among 4 mini-ferms each. And so forth ...

These are the sorts of worries that keep experimenters up at night and complicate the statistical analysis of the results (it may be a good time to seek help from a statistical expert in such cases). The point here, though, is that in order to make good tradeoffs, one must first understand what needs to be traded off. Identifying and understanding the nature of the experimental unit is an essential first step in this process. In the sections that follow, we shall see how it underlies practically everything to be discussed.

2.3   Randomization

Consider the following scenario: An assay scientist wants to experiment with a biological assay to improve its ability to measure low levels of an analyte of interest without making the assay too "noisy" – too sensitive to small changes at higher levels of the analyte. The experimental factors of interest are:

- concentration of a "capture" reagent;

- incubation time of the biological sample with the reagent;

- concentration of a signaling fluorophore

- which of two optical measuring devices is used to measure the result (two different instruments were used in the lab).

It was decided to run a so-called 2 level half-fraction factorial design. This will be show in greater detail in Chapter 3, but for now, all we need to know is that there are 8 runs in this design and each of the 4 factors is at its "+" level in 4 of the runs and its "−" level in 4 of the runs.

The 8 experimental runs were split up 4 each to be run by two laboratory technicians. For efficiency's sake, technician 1 was assigned to use one instrument and 2 was assigned the other; in this way they could work in parallel and not have to wait for each other. Another characteristic of this design is that the 4 runs assigned to each instrument turn out to be split evenly between the levels of all the other 3 factors. In particular, this means that for each instruments, there were 2 runs with long incubation times and 2 with short. Again, for efficiency's sake, the technicians agreed to run the long incubation times in the morning and the short incubation times in the afternoon. Here is a table summarizing how things were done:

Capture Conc	Incub Time	Fluor Conc	Device	Run Time	Technician
+	Long	−	A	AM	2
+	Long	+	B	AM	1
−	Long	+	A	AM	2
−	Long	−	B	AM	1
−	Short	+	B	PM	1
−	Short	−	A	PM	2
+	Short	+	A	PM	2
+	Short	−	B	PM	1

Now here's the problem. Suppose that unbeknownst to the experimenters, the two technicians have somewhat different techniques for diluting and preparing samples for measurement, and this affects sensitivity at low analyte concentrations. Here are three possibilities that might then occur:

1. The instruments make no difference, either by themselves or in combination with changes made in other experimental variables. Then the change in sensitivity would be ascribed to the instruments, a bogus, irreproducible result.

2. The instruments do make a difference, but in the opposite direction of the technicians: the technician effect would tend to mask the difference. Thus, the effect that the experiment was designed to detect would be missed.

3. The instruments make a difference, and it is enhanced by the technician effect (an interaction), so that one might conclude that the instrument has a larger effect on the sensitivity than would be seen in practice. Alternatively, the combined instrument/technician effect causes problems at high analyte concentrations, perhaps leading to the conclusion that the lower sensitivity instrument should be used in preference.

There are undoubtedly other scenarios, but the point is that the instrument/technician confusion – a technical term for this that is often used is *confounding* – can lead to spurious results. Moreover, no statistical analysis can recover the truth; the error is baked into the way the experiment was conducted.

But that is not all that's problematic here: incubation time is also confounded with the morning/afternoon time of day. While time of day itself is rarely a cause of problems, it is often a surrogate for other things that change, such as ambient temperature or humidity, amount of time a sample sits outside a freezer or since a reagent was prepared, and so forth. Such potential "lurking variables" in an experimental environment can waylay an investigator, even one who has thought long and hard about what might happen and taken care to protect against them. That is why experimental results, especially when they are unexpected or extreme in some way, must be repeated and verified.

Nevertheless, re-running experiments is expensive and time-consuming, and thus to be avoided if possible. Another line of defense against unanticipated extra-experimental effects – a sort of insurance policy against them – is *randomization*, defined as:

> **Randomization** Random assignment of experimental treatments to experimental units; if experimental runs are done sequentially, random assignment of the order in which the runs are done.

The underlying idea is simple enough: instead of doing things "systematically" – in time, in assigning equipment, people, raw materials, and so forth to the experimental units – mix such assignments up as much as possible. This acts as a kind of "insurance policy" against unknown but systematic influences distorting the results. So, for example, in the scenario above, have both technicians use both instruments and run the long incubation times in both morning and afternoon.

The question then becomes, *how* should things be "mixed up"? The "random" in the definition refers to one way to do it: use some kind of random procedure to do this mixing up (the details typically depend on the situation) . For example, one could label the eight experimental runs in the example as "A" to "H," write these letters on 8 slips of paper, throw the slips in a box, shake the box up (this is known as *physical* randomization), and then randomly pick them out one at a time. You would then run the 8 experimental runs in the order their labels were draw. In other words, if "B" were the first label picked, the "B" run would be run first; if "G" were second, the "G" run would be the second; and so forth. Of course, these days, there's no need to resort to such an actual "physical" randomization procedure: computers can do the shaking and picking.[3] Similarly, one could have the computer randomly sort a sequence of 4 "1's" and 4 "2's". If the randomly sorted order came out 21122211, then the first experimental run would be done by operator 2, the next by operator 1, the next by 1 again, and so forth. So now both the order of the runs and which operator runs which run, including with which instrument, has been randomly mixed up.

Incidentally, randomization – and also *blinding*, another important matter that we discuss in the next section – should also extend to the evaluation/measurement phase of the experimental process. Measurement can be considered a manufacturing process that manufactures numbers, or whole collections of numbers, such as a graph or image. Even human evaluations, such as rating the clarity of a plastic or the taste of a food product, are kinds of measurements. Like any other "manufacturing" process, measurement processes are themselves subject to various external, uncontrolled influences that can affect the measured result. We hope that these effects are inconsequential, but there is no guarantee. So just as we don't want extra-experimental influences to bias experimental results, we also do not want such influences during measurement to bias the assessments of those results. Randomization in the measurement process – in the order in which things are measured, in who does the measurements, and so forth – is therefore good insurance to consider there for exactly the same reasons as in the experimental process.

There are, however, at least two problems with randomization that even this simple example reveals. First, doing things in a random order can decrease efficiency and increase complexity. These are not mere minor annoyances. Experimental time and resources are typically subject to tight constraints – it is not inconceivable that randomizing the experiment would violate the contraints and prevent it from being

---

[3]Strictly speaking, computers cannot algorithmically generate true "random" numbers. By definition, an algorithm cannot produce random results. Instead, computer algorithms that produce "pseudo"- random numbers are used. That is, they generate (often lots of, for simulations) random numbers that behave "similarly enough" to random to be considered so. Determining algorithms to do this and defining "similarly enough" is a special, complex branch of computer science.

conducted! Even if this is not the case, any additional complexity invites error by making it harder to know what has to be done, when, and by whom. Consequently, implementing effective randomization can requires extra care and attention to detail to avoids such errors.

Second, and perhaps a bit more subtly, if the goal is to avoid having extra-experimental influences like time of day something is run or the identity of the person running it bias the conclusions, it would seem that one could do better then merely leave these matters to chance. What if, for example, the order in which the slips were drawn from the box (physically or virtually) were A-B-C-D-E-F-G-H? It's possible after all (1/40320 chance). Would any sensible experimenter really accept this order even though it was given by a perfectly correct random process?

Perhaps even more to the point, if one is concerned about avoiding confusion with a systematic morning vs. afternoon difference or a possible operator effect, could one not arrange to conduct the experiment in a way that deliberately avoids it?

The answer to this is yes – that is the subject of *blocking*, which will be taken up in Section 2.5. There is an important distinction to be made here, however. Blocking can only be done when the specific extra-experimental influences like time or technician are identified beforehand, but the whole point of randomization is that it is meant to protect against *unknown* external effects. This is often summarized in the dictum: Block against known sources of error and randomize against the unknown.

The obvious question that arises from this discussion is how does one decide what is the proper tradeoff between the costs of randomization – e.g. the inefficiencies and increased complexity – and the protection against external influences that it buys. Unfortunately, just as with any other form of insurance, there is no simple answer. The best advice that can be given is probably the same that one receives for home or life or health insurance: buy as much as you can afford – randomize as much as you think you reasonably can.

2.4 Blinding

Blinding The practice of hiding or masking the identity of experimental treatments from those performing the experiment or evaluating experimental results. The intent is to avoid bias by those involved in the experimental study.

Note that this is a psychological, not a statistical, precaution. For example, it is standard practice in the clinical trials of new prescription drugs, the legally mandated testing required to show that they are safe and efficacious. In a trial that, for example, compares the effect of a new drug with that of the "standard of care" of an existing drug, if physicians or patients knew which drug they were receiving, it might change their perceptions or even their physical responses. Most people believe that newer is better, and this alone might result in an improved response, even for diseases (like cancer) where this seems completely unlikely. This is an example of the so-called *placebo effect*.[4]

Of course, these sorts of issues don't arise in most scientific experimentation. Nevertheless, people often have conscious or unconscious biases that could effect experimental results. In some situations, there may be a cost incentive if results come out a certain way, e.g. a comparison of two sources of raw materials in a manufactured product or or reagents used in a laboratory; or promotion, publications, or professional standing – e.g., the validation of new theory – might be at stake. Scientific deception is not the issue here. It is rather the subtle and typically unconscious biases that might creep in.

As an example, suppose an experiment is to be conducted to improve the abrasion resistance of a plastic coating. The experimental protocol consists of applying different formulations of the coating to standard test panels with a manual sprayer, letting them dry, and then measuring the abrasion resistance by manually sandblasting the coating for a fixed time period and measuring the thickness of the coating that remains.

Even though the procedures for both coating application and sandblasting are undoubtedly standardized, it is still almost certainly the

---

[4]There is a large literature on this, but see e.g.
http://en.wikipedia.org/wiki/Placebo

case that technique matters, and that the possibility of unconscious bias therefore exists. Of course, it would be better if both application and abrasion were carried out robotically. A robot follows its programming, and since the exact same programming would be used throughout, it doesn't matter whether the robot – or its masters! – knows what the treatment is. But if this is not possible, it would be better to blind individuals involved in both the coating application and abrasion to treatment. Then randomization of the treatment order would reduce or eliminate the risk of their unconscious biases influencing results.

Similarly, if the final measurement is done manually, it too should be blinded (and randomized in any case). Generally, the physical procedure to do such blinding is fairly simple: A master list associating codes with treatments is generated and the treatments are prepared and labeled with these codes by individuals who are not involved with the testing or evaluation. All sample units use the codes as treatment identifiers, and all evaluations are recorded by these codes. The data are only unblinded at the end when final results are available and the data analysis is conducted.

There are, of course, many situations where blinding is not possible, too cumbersome, or even potentially dangerous. For example, if the addition of an additive to the coating is one of the factors being tested and presence of the additive changes the coating from clear to cloudy, this would clearly make blinding to the presence of the additive impossible. Or if the additive changed the appearance of the dried coating, this would be a tip-off to those doing the evaluation. If some of the treatments required special handling (e.g. for environmental hazards) while others did not, this might also make blinding – e.g. by using special handling for all treatments, whether required or not – too difficult.

Sometimes blinding (and, to some extent, randomization) should not be done. Suppose, for example, that too much additive in the coating might cause it too "slump" and not coat properly, and experimenters were uncertain about how much this would be. In the multifactor experiments discussed in this handbook, approximately half the coating runs would contain the additive. It might therefore be wise to know which they were. That way, the first time such a run were done, if the coating slumped, the amount of additive could be lowered so as not too waste a lot of effort and expense on clearly bad coatings. We shall discuss this idea more generally in Chapter 3.

In summary, blinding experimenters to treatment when an experiment is performed or evaluated can often help avoid unconscious biases. While in many situations it may be irrelevant or inappropriate, when this may not be the case, it should be considered.

2.5 Blocking and Split-Plotting

Blocking and *split-plotting* are related ideas that most experimenters are already familiar with, although they may not know the terms. It is difficut to define them formally without getting lost in mathematical notation, but the basic idea is that there are physical constraints on how the experiment is (or must be) conducted that cause the experimental runs to be grouped in certain ways. This grouping, which can be thought of as restrictions or limitations on how randomization can be done, may cause results within a group to be more similar then results between groups. Another way to think about it is that the grouping introduces one or more additional levels of experimental unit – the levels associated with each separate run within a group and the levels associated with the grouping(s). Unless care is taken, this can again introduce bias and result in spurious conclusions.

We use a series of examples to illustrate what we mean.

Example 3: Blocking On Time

A manufacturer of fiberglass sheeting wants to compare 3 versions in development to see which is most impact resistant. The small scale pilot facility in which the material is made can produce 2 batches per day, so that two days would be required to make samples of all 3 versions, say versions A and B the first day, and version C the second.

However, several project team members voiced concerns about this protocol, pointing out that the pilot facility must go through a startup procedure each day, and that there are often day to day differences in results related to this. This would mean that any difference seen between version C and the other 2 versions might be due to startup effects, not the versions.

It was therefore suggested that the version A be rerun on the second day with version C. In this way, version A would act as a sort of "control" – subtracting A's value from B on the first day and C on the second "normalizes" them for any day to day shift. For example, if both results on day 2 were 10 higher than they would be on day 1, then subtracting A from C gives the result that C would be on day 1 without the shift. These "normalized" values of B and C can then be directly compared.

Mathematically, this can be expressed as:

1. Day 1 normalized value of B: $[B - A]_1$ (the subscript indicates the day)

2. Day 2 normalized value of C: $[C - A]_2$

3. Normalized B/C comparison = difference between these: $[B - A]_1 - [C - A]_2$

However, while this is certainly correct in the sense that a possible day-to-day shift does not enter into and confuse the normalized comparisons, there is a certain "imbalance" to this approach: the comparisons of B and C with A involve only one calculated difference, while that between B and C involved the difference of two. Another way of saying this is that version A was run twice, while B and C were run only once. What makes A special? – why not run B or C twice instead?

Although the above practice of "normalization" is widely used, especially in biology, it turns out that the imbalance in this way of doing things actually could possibly be an issue. One way of dealing with it is to add a third day in which versions B and C are run. Note the symmetry here: there are three possible pairs of the two versions, AB, AC, and BC, and each pair is run on one day so that all three versions are run twice each.

Then the B/C comparison could be formed as the average of that given in (3) above with the direct B − C difference from day 3. That is, the normalized B/C comparison is the average of $[B - A]_1 - [C - A]_2$ and $[B - C]_3$. Algebraically (recalling that an average two values is simply their sum divided by 2) this becomes:

$$B - C = \frac{[B - A]_1 - [C - A]_2 + [B - C]_3}{2}$$

Note again that this way of doing things "blocks out" (hint hint!) any possible day to day shift – they simply do not affect the result when it is calculated this way.

Perhaps even more interesting, because of the symmetry, the A/B and A/C comparisons can now also be calculated in exactly the same way as:

$$B - A = \frac{[B - A]_1 + [C - A]_2 + [B - C]_3}{2}$$

and

$$C - A = \frac{[B - A]_1 + [C - A]_2 - [B - C]_3}{2}$$

In other words, all three possible comparisons are formed by adding and subtracting all the results, just in different ways. This is another manifestation of the theme stated in the introduction – DOE gets the most information possible from the data. And because all comparisons are from within day results only, any possible day-to-day shifts do not affect them.

This is an example of blocking in which the days can be considered blocks: results are more likely to be similar within days – they may be all be shifted up or down together – then between. Or, as we indicated at the beginning of this section, there are two levels of experimental unit, not one: days *and* individual version runs within each day.

Formally, this design is known as a *balanced incomplete block design, or BIBD*. Admittedly, this is a mouthful, but as we can see, it involves issues experimenters routinely deal with, e.g. via controls and normalization. Using controls has the advantage of being both straightforward and (in this case, anyway) requiring fewer runs. However, it can have some disadvantages. For example, suppose that the version B result on day 1 was somewhat anomalous. Then it will affect both the A/B and A/C comparisons. The BIBD would be less affected, because the version B results on day 3 have a chance to "dilute" the anomalous result.

Actually, this is just another way of saying the 2-day (= 2 block) experiment is a one versus one comparison of the kind described in the fermentation experiment of Section2.2: version B and C are run only once each, and there is no information available on the within day run-to-run experimental variability against which to judge the changes that are seen. The BIBD does have this information[5] and, because averaging is used for the comparisons, one can get a slightly more precise assessment of the results.

These statistical niceties are not our main concern here, however. The important point is that the days are an example of experimental blocks, and both experimental strategies are designed to prevent these possible block effects from corrupting the assessments of the effect of the experimental factor of interest, the fiberglass version.

Example 4: Blocking on Plates in an Experiment to Improve a Biological Assay

Biochemical assays of many types are routinely used in medical testing of blood, urine, and other kinds of human tissue samples by medical laboratories. The testing is often done in multiwell *microtiter* plates on automated equipment to improve throughput and accuracy. A microtiter plate is a rectangular, flat, usually plastic plate with multiple wells arranged in a rectangular array that act as individual test tubes in which biochemical reactions occur. Here is a picture of a 96 well plate arranged in an 8 row x 12 column array, probably the most

---

[5]but only minimally

widely used format at present.[6]

Figure 2.1: A 96 Well Microtiter Plate

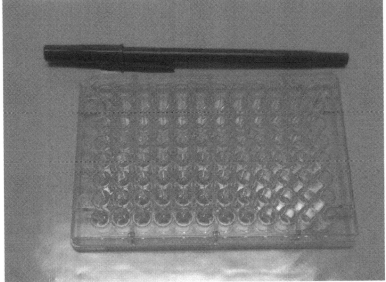

There are typically many components of a biochemical assay that must be carefully balanced to assure that the assay has the necessary sensitivity, precision, accuracy and robustness (among other requirements). This can be challenging; considerable expertise and experimentation is frequently necessary, and multifactor DOE is often used as part of the assay development process.

Suppose that six separate solutions of an analyte have been prepared in which various kinds of contaminants were added as a way to determine how well an assay to measure the analyte works in the presence of the contaminants. This is an example of what is often called an assay *robustness* study. Each contaminated solution is split into two parts – *aliquots* is the technical term – one to be measured by the existing assay and one to be measured by a new assay that uses different

---

[6]However, 32 x 48 plates of the same size, 1536 wells! (and more), are used in some labs.

reagents.

An assay "measurement" consists of a *serial dilution*. Here is an explanation of what this means.

Given the assay *dilution factor* of k and the number of points, n, in the dilution series (including no dilution), then from an undiluted sample with analyte concentration, C, n dilutions will be made with concentrations: C(no dilution), C/k, C/k^2, C/k^3, ... , C/k^{n-1}. These are done serially – each subsequent dilution dilutes the previous one by a factor of k. As a numerical example, suppose $n = 10$ and $k = 2$, with C = 200 (in some appropriate units). Then the concentrations are 200, 200/2 = 100, $200/2^2$ = 100/2 = 50, $200/2^3$ = 50/2 = 25, ... , $200/2^9$ = .39 .

For each dilution, the measured response is typically an optical or radiation intensity. So the response for the whole series consists of the set of intensities corresponding to the dilutions, typically shown as a graph of response versus dilution factor, 0 to n. What is important to know here is that for practical reasons the dilution series for each sample is almost always laid out either in a single row or column of the plate. Suppose in this case it's columns, and so an 8 point serial dilution. With 12 columns on the plate, this means that 12 samples/runs can be run on each plate.[7]

Suppose, further, that it was decided that two replicates (see Section 2.6) of each sample needed to be run. This means that 24 samples – the six kinds of contamination $\times 2$ =12 runs per reagent – need to be run. With 12 columns available on a plate, this means that at least two plates are required.

Now there are three obvious alternatives how this might be done. First, the 24 samples could be completely randomly assigned to the 24 plate columns; each plate could run the 12 replicated contaminations with a single set of reagents; or each plate could run a single replicate of all 6 contaminants with both reagents. Examples of each, where A/B indicates the reagents used, 1 - 6 indicates the contaminant type, and the assignment to the 12 columns are given in order:

---

[7]Actually, edge columns or rows are often left empty or used for controls, but we ignore this complication here.

Completely Randomized

Plate 1: B5, A3, B4, A3, A2, B3, B3, A4, B6, B4, B5, A5

Plate 2: B2, A1, A1, A2, A5, A4, B1, B1, A6, B2, A6, B6

Note that 5 of the "A"-type assays fell into Plate 1 and 7 in Plate 2; and conversely for the "B"-type, of course.

Single Assay Type in a Plate

Plate 1: A5, A6, A1, A2, A3, A2, A4, A1, A6, A5, A3, A4

Plate 2: B1, B5, B2, B6, B4, B3, B3, B1, B2, B5, B4, B6

Note that the column order of the 12 samples within a plate has been randomized.

All 6 Contaminations for Each Assay Type Per Plate

Plate 1: A2, B5, B1, A1, B4, A3, A4, A5, B3, B6, B2, A6

Plate 2: A6, A3, B6, B3, B2, A2, B4, A4, A5, A1, B1, B5

Each sample contamination type is tested with both assay types in each plate with the column order randomized.

So which of these three designs is the "best"? To some extent, it may depend on the difficulty of actually doing things as specified. For example, if the experiment is performed manually, it is probably more difficult to do things in the totally randomized order of the completely randomized design than either of the other two; and using only a single reagent type per plate as in the second design is probably simpler than using both assay types as in the third design. On the other hand, it may make no difference if it's an automated assay requiring only some minor programming differences.

However, as all those who work with such assays know, there is an important reason that the third design would be preferred: there are often plate to plate shifts in how the assays perform. The purpose of the experiment is to compare the A and B assay types. In the second design, all the "A" samples are tested in the first plate and all the "B" samples in the second. As a result, there is no way to know whether any systematic shift seen (or not, for that matter) is due to a difference in the plates or the assays. The completely randomized design (by chance) largely alleviates this problem, because both A and B assays occur in both plates. But there is a slight imbalance, as noted. By deliberately testing one replicate each of all 6 kinds of contamination for both assays in each plate, the assay type comparisons only involve within plate differences. In other words, treating plate as a block, design 3 blocks on plate and thereby eliminates plate to plate differences, exactly as within day comparisons did in the fiberglass experiment.

This is, in fact, an example of a so-called *randomized complete block design* (it is "complete" because all the treatments occur in each block). It is perhaps worth noting that by using such a design it is not necessary to use controls on each plate to normalize plate values, although this is widely done(and can be useful for other reasons). In addition, we note that randomizing the assignment of columns to samples within each plate might also not be the best strategy if there are systematic *positional trends* within a plate. This can sometimes happen due to the optics of plate readers, how the plates are made, or how the reagents are pipetted into the wells. To deal with this possibility, it might be better to block on position within a plate instead of randomizing, in addition to blocking on the plates. Other then raising the possibility, however, we shall not discuss it further. It's the sort of thing that needs to be taken up with local statistical experts if it's a matter of concern.

Example 5: Split Plotting With Microtiter Plates

Suppose that in the previous assay example, in addition to comparing the different sets of reagents = assay types, it was desired to investigate the effect of incubation time, the time that reagents and samples are allowed to react until the final activity measurement is made. This is always an important consideration – too little time and the reactions do not have sufficient time to complete; too much and throughput is adversely impacted.

However, there's a difficulty: only whole plates can receive different incubation times, so that all wells in a plate must be incubated for the same amount of time. This means that there are two different sample units at play here. For the assay reagents, the well – or the plate column for a serial dilution – is the sample unit; but for incubation time, it is the whole plate. In other words, any assessment of the effect of changing the incubation *must* involve a comparison of results between plates. But we already know that plate to plate differences may exist simply because of random plate to plate differences unrelated to any deliberate changes made between the plates, That is why we blocked on plates in Example 4. So how can one assess the effect of changing incubation time (if any) without possible plate effects confusing matters?

The standard, tried and true answer to this question is to use controls on each plate and normalize all results to them. Unfortunately, this can't work: the plate is still the sample unit for incubation, meaning that normalization removes any incubation time effect (which is confounded with random plate to plate differences)!

In order to determine the effect of an experimental factor such as incubation time that has a different experimental unit (plate) from that of the other factor, reagent (well) it is necessary to replicate that unit – plates, in this case. For example, if the third experimental plan were used(all 6 samples assayed with both sets of assay reagents in each plate, twice), the whole experiment would need to be replicated, requiring 4 plates in all instead of 2. Moreover, the analysis for the incubation time is different than that for the reagents – there is more information available to assess the reagent effect then the incubation time effect, and the experimental variability in replicating plates is likely to be different (probably larger) than replicating treatments in wells within a plate. The analysis needs to account for both of these issues.

These kinds of experiments in which there are one or more different levels of experimental unit for one or more different experimental factors are known as *split-plot* experiments.[8] The plates are the *whole plots*

---

[8]The terminology derives from agricultural experimentation, where the experimental units were actually plots in a field.

in this experiment, while the columns within plates are the *subplots*. As in this example, in such experiments it is necessary to replicate the whole plots in order to assess their effect; and even so, the whole plot treatments are usually assessed less precisely (= less information is available) than the subplot treatments. While we do not wish to get mired in the statistical swamp, the take home message is that any analysis of the data from such experiments that does not recognize this may produce spurious conclusions.

Example 6: Split Plotting With Easy-to-Change and Hard-to-Change Factors

As another example of a common but somewhat subtle situation that leads to split plotting, consider an experiment whose goal is to improve the quality in a flat screen display coating process. Suppose two factors are being studied, the temperature of the coating material and the rate at which it is applied. Suppose 2 test panels/runs will be made at all 9 treatment combinations of 3 rates and 3 temperatures. It turns out that it is easy to change the rate – essentially, just turning a dial changes it. Temperature, however, is much harder, as the reservoir containing the coating material has to be be heated or cooled and allowed to equilibriate. This can take a considerable amount of time.

Now if the order of the 9 temperatures is completely randomized, then the temperature might have to be changed frequently, perhaps even for every run – 18 times – if the experimenters got very unlucky. To avoid this and minimize experimental time, it was decided to run the first 6 runs at low temperature, the next 6 at the mid temperature, and the last 6 at the high temperature. Within each group of 6, the 3 rates would be run in random order.

This should sound familiar. It is similar to Example 3 in which the blocks were days, that is, units of time. In the present case, the time over which the experiment takes place has been grouped – blocked – into three separate periods of time by specifying that the temperature must change systematically instead of randomly over time. Of course, in the previous example the blocking was explicit because of known possible day-to-day startup effects that could shift results. That is not the case here. Nevertheless, by systematically increasing the temperatures over the course of the experiment and restricting the number of temperature changes, it makes it possible for any extra-experimental time effects and/or variation in the coating temperature from its nominal level to get confounded with the temperature changes in exactly the same way as day-to-day shifts could get confounded with the assay reagent changes. As we argued in Section 2.3, randomization is the insurance that one buys precisely to avoid this sort of problem. Foregoing it increases the risk of spurious conclusions.

Again, one can think of the experiment has having two levels of experimental unit, the individual panels within each time period and the groups of 6 panels within each temperature (because temperature is only changed between groups). Strictly speaking, temperature is confounded with the possible random group to group differences, so there is no way to know whether what looks like a temperature effect is really nothing more than random group to group variation caused by something else that is changing over time.

To avoid this confounding, one would have to break at least some of the groups into separate subgroups in which temperature was reset. For example, instead of having 3 groups of 6 panels run without resetting temperature, one could have 6 groups of 3 panels where temperature is reset for each group. One would then have a "within" group (i.e. panel to panel) level of variability to assess the coating changes, and a "between" group (group to group) level to assess the temperature changes. This is rarely done, of course. The pain is too great and the analysis too complicated. Instead, the temperature grouping is ignored and the data are analyzed as if the deliberately specfied run order were random and the temperature reset at every run. That is, we pretend that there is only one level of experimental unit, the individual panel.[9]

Most of the time this sort of "cheating" is probably harmless, be-

---

[9]And for an even more confusing possibility, suppose the 6 panels within a temperature group were run as 3 pairs in which the coating was held fixed for each pair. Now there would be 3 coating subgroups within the 3 temperature groups. In theory the coating effect at each temperature would be confounded with these subgroups and the coating effects at the different temperatures could not be rigorously assessed.

cause there is no additional variability because of the grouping over time. But the point is again that a lack of awareness of the statistical consequences of even the apparently sensible practice of restricting the number of changes of hard-to-change factors in the cause of experimental efficiency can lead to surprising – and unwelcome – consequences. It is not that experimental efficiency must be sacrificed at the altar of statistical correctness: sometimes factor changes *must* be restricted in this way. Rather, the point here is that understanding the issues should lead to greater caution in interpreting results for the whole plot, hard-to-change factors than for the within plot easy-to-change ones. And if the experimental stakes are hign enough to warrant it, explicitly running such experiments as split plot designs (e.g. by replicating whole plot treatments) can sometimes permit sufficient statistical rigor *without* too great a hit to experimental efficiency. In such circumstances, consult your local statistical expert for help.

2.6   Replication

Readers may be surprised that we have saved this issue for last. After all, statisticians are often accused of advocating more replication and bigger sample sizes than is reasonable or sometimes even possible. So why have we waited until now to discuss what seems to be so central a statistical concern?

Our answer is simple: we think (explicit) replication is vastly overrated, and in the multifactor designs that are the subject of this book, largely a waste of effort.

This view clearly requires justification. To begin, we first need to formally define what we mean by the term.

> **Replication** Separate experimental runs in which all factors have the same settings. That is, experimental replicates are different experimental runs = different experimental units for which the experimental treatments are identical..

Before we explain why we think experimental replicates aren't generally worth the effort, it's probably worthwhile to ask why they might be useful in the first place. After all, one might naively assume that because all replicates receive exactly the same experimental treatment, they should all produce exactly the same experimental result. In practice, as every experimenter knows, this doesn't happen due to experimental and measurement variability: separate experimental runs (should) use separate experimental units, receive separate sample preparation and setup, and so forth. All this will cause results to vary.

But that is exactly the point! – because replicate results differ due only to experimental variability, they provide a way to measure how much variability there is. Such assessments enable one to distinguish between what is "merely" variability and what is likely "real." Formal statistical implementation of this idea is known as *statistical inference*, which is widely regarded as an objective way to draw scientifically sound conclusions from data.

But there are two, typically fatal, flaws to this reasoning.

1. As we mentioned at the end of the Introduction, the experiments that are the concern of this handbook are *exploratory*, not *confirmatory*. They should be viewed as part of an overall scientific learning process. In particular, they are not designed to test pre-specified hypotheses, a necessary condition for (standard) statistical inference methods to be applicable. Moreover, they are generally too small to provide the information on variability needed for statistical inference to yield useful conclusions in any case.

2. More subtly, and perhaps of greater concern, the information about variability gained from experimental replicates is likely to be wrong, and using it as the basis for inference can lead to wrong conclusions.

Here's what we mean by this. The essence of all statistical inference is to form some sort of signal-to-noise (S/N) ratio, using statistical theory to determine whether the ratio is sufficiently large to indicate that the "signal" is greater than what would would be expected if the signal really wasn't there but was just noise, itself. The details here aren't important; what is important is that the "noise" part of this ratio should be an estimate from the data of all the experimental variability that occurs from run to run in the experiment, *whether the runs are the same or different treatments (= combinations of factor settings).* Practically speaking, what this means is that in order to get an "honest" estimate of noise for inference, one must go through exactly the same procedures when doing replicate runs – including separate setups, sample preps, equipment adjustments, etc. – as one would do if the runs were completely different.

This is closely related to the split plotting issues discussed in the previous section and to the examples in Section 2.2 on experimental units. The common thread is that some steps in the experimental procedure (using separate sample units, changing factor levels) are skipped so that the run to run variability is reduced within certain groups of runs (those with the same experimental unit, those with the same whole plot factor setting) and increased between the groups.

The same often happens with replicates: for reasons of experimental convenience and efficiency, all steps are not repeated among the replicates, so that the experimental variability among replicate runs is less than that between non-replicate runs. We call such "pseudo"-replicates *bang-bang* replicates, to indicate that they are done quickly and easily compared to runs where the factor settings change. Bang-bang replication tends to make the estimate of experimental noise in the S/N ratio too small and hence the ratio too large. This results in *false positives*, conclusions that experimental factors have larger and more meaningful effects then they really do. This, in turn, can cause *irreproducibility*: the purported effects cannot be reproduced either in future or by others. This is a bad thing.

Of course, this need not be the case. One can run replicates in which all steps are properly performed and the experimental variability is the same among the replicates as between different treatments. But our experience is that this is not how things are done most of the time. The temptation to shortcut procedures is just too great.

It turns out that, with suitable caveats, one *can* actually get a trustworthy estimate of experimental variability without running replicates. This relies on the so-called *hidden replication* built into the designs, a matter to be discussed in Chapter 3. However, even in this case, statistical inference based on such estimates is not generally reliable. The need for efficiency due to cost, time, and resource constraints generally requires designs that are too small.

That is the origin of our claim that replication is not typically useful. If it's easy to do, it's probably not providing proper estimates of variability. If it's hard – if careful replication is being done – it's probably more useful to expend the effort using a different design that provides more information. The hidden replication will provide (about) the same information on experimental variability, anyway.

The one exception to this advice against replication – to be discussed in the next chapter – is that when center points are run, it is generally useful to replicate them (properly! – not bang-bang replicates) a few times. While this still will not provide sufficient information for useful statistical inference, it can serve as a kind of overall quality check on the conduct of the experiment. If the variability among replicates is uncomfortably large, it may indicate a need to modify experimental procedures or even scale back experimental objectives. This is important to know.

2.7 Chapter Summary

This chapter has described some key themes of experimental design and execution. While statistical matters are certainly one component of what's involved, we believe that the underlying issues are both broader and more practical than "mere" statistical methodology. Determining the experimental unit in order to make correct assessments of treatment effects, using randomization and blinding as "insurance" against experimental bias, employing blocking, split plotting, and proper replication to reduce systematic extra-experimental effects and obtain better information on the effects of interest – these are basic requirements for achieving reliable, reproducible conclusions from *any* experimental results, whether formally "statistically designed" or not.

As we indicated, a complete formal treatment of these issues can get very involved, occupying hundreds of pages. Our goal is only to provide a non-mathematical conceptual understanding sufficient to make experimenters aware of their potential impact on their results and when to seek expert statistical advice.

Having said that, however, we believe that most of the time experimenters can avoid statistical complexities and handle these matters perfectly well on their own. The most important expertise for experimental design is subject matter knowledge and experience, clear understanding of what may be important, of what can go wrong, and of what is necessary to get trustworthy data. Our view is that effective experimentation and robust conclusions are more likely when this expertise is informed by an understanding of the ideas we have presented here. We hope that you agree.

Chapter 3

Design

The purpose of this chapter is to provide a catalogue of 2-level experimental designs with instructions for their selection and use. As will shortly be explained, the organizing principle underlying this catalogue is the property of *design projectability*. We have found this to be both useful and easy to understand, and we hope that it helps to make multifactor experimentation more accessible to practitioners.

In the appendix to this chapter, we provide both design tables in printed form and instructions to find links to equivalent tables on the Web. For all but the smallest designs, as it is both tedious and error prone to deal with them manually, you should download the designs[1].

3.1 Preliminaries

3.1.1 Why 2-level Designs? – The "Curse of Dimensionality"

In the Introduction, a biofuels experiment with 2 factors was described, but it is not uncommon in industrial applications – in process validation and improvement, for example – to have processes with as many as

---

[1]Except for the Hall H2, for which we could find no convenient link

dozens of factors that, in theory anyway, could be controlled and ma-
nipulated. Designs with relatively few runs that could handle so many
factors are actually readily available. The real limitation is practical,
not statistical: it is impossible to accurately control and manipulate
"too many" experimental factors at a time. Five to ten is typical,
although some designs in the catalogue allow for considerably more if
needed.

One problem that must be confronted even with fewer factors is what
is colorfully referred to as "the curse of dimensionality." Although
images of Dracula may come to mind, the reference is actually more
prosaic, and can be easily explained with an example or two.

Suppose first that we have a 2 factor experiment with factor A having
3 levels, a1, a2, and a3; and factor B with 2 levels, b1, and b2. Then
there are 3 x 2 = 6 possible combinations of these two factors:

a1 b1
a2 b1
a3 b1
a1 b2
a2 b2
a3 b2

Similarly, if there were a 3rd experimental factor, C with 4 levels, c1,
c2, c3, and c4, then there would be 3 x 2 x 4 = 24 possible combinations:

a1 b1 c1
a2 b1 c1
a3 b1 c1
a1 b2 c1
a2 b2 c1
a3 b2 c1

a1 b1 c2
a2 b1 c2
a3 b1 c2
a1 b2 c2
a2 b2 c2
a3 b2 c2

a1 b1 c3
a2 b1 c3
a3 b1 c3
a1 b2 c3
a2 b2 c3
a3 b2 c3

a1 b1 c4
a2 b1 c4
a3 b1 c4
a1 b2 c4
a2 b2 c4
a3 b2 c4

The general pattern is obvious: If you have m factors, F1, F2, ... , Fm with numbers of levels k1, k2, ... , km each respectively, then there are k1 × k2 × ... × km combined possible runs in all. Note that if the number of levels is the same for each factor, say k1 = k2 = ... = km = k, then the product of these is just k × k × ... × k (m times) = k^m. The term "curse of dimensionality" derives from viewing each factor as a "dimension" in space[2]. While more than 3 dimensions might seem strange, for our purposes, it's not much different than 2 or 3 dimensions, and thinking in these terms should suffice. The "curse" part comes because of the "exponential growth," k^m, in the number of runs: if k − 2, a three factor experiment has $2^3 = 8$ possible runs; 4 factors has $2^4 = 16$; 5 factors has $2^5 = 32$, and so on. With 3 levels it gets out of hand even faster: 3 factors already has $3^3 = 27$ possible runs, 4 factors has $3^4 = 81$, and 5 factors has $3^5 = 243$! Clearly, if we want to run all possible combinations in an experiment, a so-called *full-factorial* experiment, the number of runs gets too large to be practical for more than 4 or 5 factors with 2 levels, and more than about 3 factors with 3 levels! However, it turns out that if one sticks to 2 level designs, there are ways to evade – or at least ameliorate the effects of – this curse.

What is evidently required is some way to select a subset of the possibilities in some clever way so that "most" of the important information that could be obtained by running all the possible combinations of fac-

---

[2]When the factors are continuous, anyway

tor settings is still gained, but with a drastically reduced number of
runs. That is exactly what we will show how to do! In fact, the designs
for 2 level factors given in this chapter allow up to 11 factors to be
studied in 12 runs (instead of $2^{11} = 2048$ runs); or up to 12 factors in
24 runs. In fact, some of the designs can be expanded to handle even
more factors[3], but as we've already discussed, there is rarely a need
for so many. However, the larger designs still turn out to be useful for
fewer factors, as we shall see.

3.2 The Pareto Principle

One of the two big ideas that underlie the multifactor design strategy presented here is known by various names, the Pareto principle[4], the 80-20 rule, effects sparsity, the vital few and trivial many, and the Principle of Parsimony among them. We define it here as:

> **The Pareto Principle** For processes with many possible causes of variation, adequately controlling just the few most important is all that is required to produce consistent results. Or to express it more quantitatively (but as a rough approximation, only), controlling the vital 20% of the causes achieves 80% of the desired effects.

This is nothing new or revelatory. All it says is that to get the most bang for your buck, focus on the important stuff and ignore the unimportant. Of course, the challenge is determining what's important and what isn't. It's probably not much of an exaggeration to say that figuring that out is what science and engineering are all about. But for this reason, it's also true that there's always a body of theory and practice to guide such investigations: experimentation doesn't occur in a vaccuum. Very often, that knowledge is all that's needed to identify and control what's important. But when it's not, or when we are uncertain about its adequacy, we need to experiment in order to enlist Mother Nature's help.

---

[3] The PB20 can actually handle up to 19

[4] Vilfredo Pareto was a 19th century Italian economist who provided a mathematical formulation of the idea

Naturally, all available knowledge and experience should first be used to whittle the list of experimental factors down to just those that really need to be investigated. Almost always, this will be just a few – sometimes even just one or two, and rarely many more than a handful or so. While this claim cannot be proven in any rigorous way, there are good reasons to expect it to be true. Basically, to ever get to the point of having a defined and controllable engineering or scientific process to experiment on, there cannot be "many" process variables whose effects on the outcome are not sufficiently understood – otherwise one wouldn't have gotten to a workable process in the first place!

This is where Pareto and the *projectability* concept discussed in the next section come into play. As we have seen, experimentation with two, three, or possibly four factors in 2 level designs is manageable: one can use full factorials (possibly with replicated centerpoints, Section 3.5.3) to experiment at all possible combinations. But beyond that, the curse of dimensionality rears its head, and other approaches are required. It is the Pareto Principle that makes these approaches work by assuring that identifying and controlling the most influential few – typically just two or three – will achieve the desired results. Examples that illustrate this will be given in Chapter 4.

The task for now is to show how designs with just a few runs can put Pareto to work. This is the where the second big idea, design *projectability*, plays a key role.

3.3   Projectable Designs

In the rest of this handbook, we follow the convention of the Introduction, using "−" and "+" to label the levels of the 2 level factors in the designs. These will indicate "low" and "high" settings of factors like temperature or concentration when such ordering makes sense; when it does not, it just distinguishes the two different categories for the factor, such as which reagent type or which instrument. With only 2 levels, doing things this way works for both types of factors (but does not with more!).

We now give an example that demonstrates the concept of design *projectability* that is the basis of our approach to multifactor experimentation. Suppose that there are 6 experimental factors, F1, F2, ... , F6. Consider the 12 run design in Table 3.1, where the columns give the factors and the rows are the runs.

Table 3.1: 12 Run Design for 6 Factors

Name Run #	F1	F2	F3	F4	F5	F6
1	−	+	-	+	+	+
2	+	-	+	+	+	-
3	-	+	+	+	-	-
4	+	+	+	-	-	-
5	+	+	-	-	-	+
6	+	-	-	-	+	-
7	-	-	-	+	-	-
8	-	-	+	-	-	+
9	-	+	-	-	+	-
10	+	-	-	+	-	+
11	-	-	+	-	+	+
12	+	+	+	+	+	+

There are a number of important properties to note about this design.

1. Each column (factor) of the design has six "+" and six "-" signs.

2. *Any* pair of columns has three occurrences each of the four pairs $(- -)$, $(+ -)$, $(- +)$, and $(+ +)$. For example, for F2 and F5, the pairs in order are: $(+ +)$, $(- +)$, $(+ -)$, $(+ -)$, $(+ -)$, $(- +)$, $(- -)$, $(- -)$, $(+ +)$, $(- -)$, $(- +)$, and $(+ +)$. The technical term for this is that the design is *orthogonal*. This is an important property of all the designs used in this handbook. We note it in the following definition:

   Orthogonal Designs: A 2−level design is said to be **orthogonal** if for any two columns, all four possible combinations of "−" and "+" occur an equal number of times. Note that this means the number of rows/runs in the design must be a multiple of four.

3. For **any 3 columns** of the design, all 8 possible combinations of "−" and "+" appear at least once; more precisely, four of the combinations appear exactly once, and four appear exactly twice.

   To better see what this means, choose any 3 columns, say F1, F3, and F6. Then the settings in those 3 columns are shown in Table 3.2.

Table 3.2: The 12 Run 6 Factor Design in Just F1, F3, and F6

Name Run #	F1	F3	F6
1	-	-	+
2	+	+	-
3	-	+	-
4	+	+	-
5	+	-	+
6	+	-	-
7	-	-	-
8	-	+	+
9	-	-	-
10	+	-	+
11	-	+	+
12	+	+	+

Also note that the pairs of runs 2 and 4, 5 and 10, 7 and 9, and 8 and 11 are identical. These give 4 of the 8 possible triplets of "-" and "+" signs; the remaining four appear in runs 1, 3, 6, and 12. . The same thing would happen but with different rows being paired no matter what three factors are chosen. Try it with others to convince yourself if you like.

This last property is called *3-projectability*, formally defined as:

3-Projectability A 2-level design is said to be **3-Projectable** if *any* 3 columns of the design contains all 8 possible combinations of the "−" and "+" settings of the factors at least once each. In other words, recalling that a full factorial design is one containing all combinations of levels of the individual factors, a 3-Projectable design is one in which any choice of 3 columns contains a full factorial design in those 3 columns.

Similarly, we define:

4-Projectability A 2-level design is said to be **4-Projectable** if *any* 4 columns of the design contains all 16 possible combinations of the "−" and "+" settings of the factors at least once each. In other words, recalling that a full factorial design is exactly one containing all combinations of levels of the individual factors, a 4-Projectable design is one in which any choice of 4 columns contains a full factorial design in those 4 columns.

Note also that by definition all orthogonal designs are automatically 2-projectable because they contain all four combinations of "−" and "+" settings. Hence all designs in this handbook are at least 2-projectable.

Clearly, 3-projectable designs must have at least 8 runs (they must contain all combinations of 3 factors at 2 levels each), and 4-projectable designs must have at least 16 (all combinations of 4 factors at 2 levels each) . Thus 4-projectable designs provide information on more possible factors at the cost of more runs.

Now the point of all this: why is projectability so useful? Answer: because, from the Pareto principle, only a "few" – perhaps even only one or two – of the factors in a design affect the experimental response "enough" to be of concern. So if a design is 3-projectable, say, we know that *if* we could figure out which (up to) 3 were the most important, the design would contain a full factorial in those factors – and, in fact, four of them would occur twice. This means that, in some sense, all the information possible about how those 3 factors, *whichever they were*, affect the response is present in the design.

To get a better feel for what this means, consider the 6 factor Projectable Designs example. We showed that this design is 3-projectable. Therefore, once we run the experiment, the results should tell us which are the 3 most important factors, and then Pareto tells us we can safely ignore the other 3. This will then leave us with a 12 run, 3 factor experiment. Of course, until the experiment is actually performed, we don't know *which* 3 factors; but we do know that, whichever they are, in those factors there will be 4 unique runs and 4 pairs of runs with identical, replicated settings. That is the – pretty amazing! – property of 3-projectable designs.

But this implies yet another important result. Any differences in results between the replicated pairs can come from only two sources: changes in the ignored "unimportant" factors and experimental variability. If the unimportant factors had no effect at all, the differences would be entirely due to variability. This would mean that although the design started off with 6 factors and 12 unique runs – i.e. no replication – it ended up as a 3 factor design with replicates and therefore some information on experimental variability.

The ignored factors probably don't have exactly a zero effect, so this is not quite the case. Nevertheless, Pareto tells us that most of the time, their effects are small enough that pretending they're zero generally gives a reasonable idea of what the experimental variability is, especially in such small designs. This kind of magic "pseudo" or "hidden" replication is important enough to formally define.

Pseudo/hidden-replication: A subset of runs in an experimental design that turn out to be different only in the settings of unimportant factors. That is, they are replicates in the settings of all the factors that are found to be important.

This is the reason for our claim in Chapter 2 that it is generally not worth spending a lot of effort replicating runs in an experimental design: pseudo-replicates will almost always give good enough information on experimental variability anyway for "free" without the additional cost of true replicates. In fact, as was mentioned already, it is often the case that in a 6 factor design like Table 3.1, not even three factors are needed; there may be only one or two that are actively affecting the response, meaning that the design is basically just a 1 factor design pseudo-replicated six times or a 2 factor design pseudo-replicated three times. Explicit replication beyond this is generally not going to provide much additional useful information on variability (except for replicated centerpoints, to be discussed later in this chapter).

Of course, all of this still leaves the crucial question: Once the experimental results are in, how do you determine which are the important factors and which are not? We defer this to Chapter 4, where a step-by-step (graphical) procedure for answering this question will be given. It's worth noting, however, that the procedure depends on the orthogonality of the designs. So there really is quite a bit going on here.

3.4   Design Catalogue

Having established the principles behind design construction, we can now give a catalogue of designs that should be serviceable for most circumstances where 2 level designs are used.

Table 3.3: **Design Catalogue**

		# of Runs				
		8	**12**	**16**	**20**	**24**
	4	2^{4-1}		2^4		
	5	2^{5-2}		2^{5-1}		
	6	2^{6-3}		2^{6-2}		
# of Factors	**7**	2^{7-4}		2^{7-3}		
	8		PB	2^{8-4}		
	9				PB	PBfo
	10			H2		
	11					
	12					

LEGEND

2-Projectable (2^{6-3})   3-Projectable(PB)   **4-Projectable(PBfo)**

Notes:

1. Detailed tables of all the designs are provided in the appendix to this chapter.

2. PB = "Plackett-Burman," the names of two British mathematicians who discovered these designs in 1946. They come in various sizes. The 12 and 20 run designs are used here. They can handle up to 11 and 19 factors respectively. We have stopped at 12, as this is sufficient for most practical applications.

3. PBfo = The "foldover" of the 12 Run 11 factor PB design. It is constructed by changing all signs in the 11 factor PB12 ("-" to "+" and vice-versa), stacking this on top of the original unswitched design, and then adding one more column of 12 "-" and 12 "+" signs.

4. Designs of the form 2^{n-k} are known as 2-level "fractional factorials" (ff's).

5. H2 is a design of Hall's (1961) and is Table 7C.1 of Box, Hunter, and Hunter (2005), and Table 3.8 in the appendix to this chapter. It was shown by Box and Tyssedal (2001) to be 3-projectable.

3.4.1 How many Factors? How many runs? – the information-cost tradeoff

There are two obvious features of this design catalogue. First, there appears to be little or no "cost" in additional runs to including more rather than fewer factors in a design. For example, 12 run 3-projectable PB designs are available for from 5 to 11 factors. So how does one decide what the "right" number of factors to include in an experiment is? Second, for any number of experimental factors that are chosen (except for just 3), there is a choice in designs, sometimes from as few as 8 runs to as many as 24[5]. Which should one choose? Or, perhaps

---

[5]In fact, as one might expect, there are many more designs that one could choose, but they require even more runs. We have limited the choice to 24 runs or less, because these are the most useful in practice.

more to the point, why not just choose the one with the fewest number of runs possible?

These two questions are related, of course. For both it comes down to balancing the costs and practical difficulty of doing the experiment versus what one needs to learn. Consider first the question of picking which factors to include. Suppose, for example, that engineers are developing shipping cartons for auto-injectable, drug-filled syringes, that is, syringes containing pre-measured doses of drugs for direct use by patients rather than for administration by a medical professional. Obviously, the package must protect the syringes from damage or leakage. But because they are shipped directly to consumers who may then transport them themselves (e.g. when travelling), there is no control over the handling and abuse the packages may have to endure under these conditions. Of course, they could just be wrapped in layers and layers of styrofoam and bubble wrap, but that would be far too bulky.

One might expect there to be many factors that determine whether a package has sufficient protection without being too unwieldy. The geometry of the cartons, the packing material, how many syringes to pack into a carton, whether to use molded syrofoam or a flexible wrapping for individual syringes, and so forth, all might come into play. Trimming down such a list to 3 or 4 factors to fit into an 8 run design might be difficult. But for a mere 4 extra runs, one could use a 3-projectable 12 run PB design to experiment with up to 11 such factors. If each experimental run is just a different carton configuration and the experimental protocol consisted of destructively testing (e.g. by crushing, dropping, shaking) several dozen cartons made up in a configuration, then whether more or fewer factors determine the details of the configurations probably makes little difference: each configuration will require its own customized design/build procedure no matter what. So having the ability to include more factors that determine the design of each configuration without having to build more configurations – i.e. increasing the number of runs – is very useful.

Very different considerations might determine the factors in an experiment investigating a lower-cost plating process for electroplating metal, however. Factors for that experiment might consist of things like the temperature or flow rate of the electroplating bath and concentrations of various chemical constituents. The experimental responses of interest might be corrosion resistance and cost (since reducing cost is a primary goal and different amounts of reagents would presumably have different costs). Since a change in concentration of the chemicals requires changing – and possibly discarding – the bath, minimizing the number of such changes that have to be made would be desirable. This argues for limiting the number of chemical consituent factors that are included in the experiment. If, additionally, temperature were included as an experimental factor, there would be a temptation to reduce the number of temperature changes because each change requires time for the bath to stabilize before the test run can be made. This might lead to split-plotting – running all the high temperature runs first and the low temperature ones second, for example – resulting in possible problems in assessing the temperature effect, as discussed in Section 2.5.

So, not surprisingly, the right number of factors to include very much depends on the nature of the experiment and the costs of including them. The point is that the DOE design catalogue allows the flexibility to do what is needed.

By the same token, the catalogue allows flexibility in the choice of the number of runs. At first glance, this may seem almost a non-issue: for any given list of experimental factors, why not just choose the experiment with the fewest number of runs possible? Why do extra work if you don't have to?

But, unsurprisingly, there is something to be gained by the extra work – increased projectability. Consider a 6 factor experiment, for example. One could choose an 8 run 2-projectable, 12 or 16 run 3-projectable, or a 24 run 4-projectable design. If it is believed that at most 2 of the 6 factors are likely to have enough of an effect to matter – that is, to be *active* – then 2-projectability is good enough and the 8-run design should be a good choice. If 3 or less, than either the 12 or 16 run 3-projectable designs would work (but which? – more on this in a moment). For those who are more risk averse and want to allow for up to 4 important factors among the 6, the 24 run design should be chosen.

There is also something else going on here. Based on the previous discussion of the curse of dimensionality in section 3.1.1, it is obvious that some shortcuts must be taken to include many factors in a design with few runs. The Pareto Principle and projectability come to the rescue most of the time, but they are not infallible: it is always possible that the experiment will not lead to clear conclusions or, even worse, that wrong conclusions will be drawn. While the details of why this might occur need not concern us here, what is important to know is something that common sense already suggests: the more experimental factors one includes in a fixed number of runs, the greater the risk for inconclusive or incorrect results. Or to put it the other way, for a fixed number of factors, choosing a design with more runs generally provides more information and therefore reduces this risk.

That is why a 16 instead of a 12 run design for 6 factors might be chosen even though both are identically 3-projectable. A natural question to ask is how much does this help? – how much is the risk reduced for the cost of 4 extra runs? Unfortunately, this is difficult to determine, as it depends on both the usually unknown level of experimental noise and the certainly unknown sizes of the various factor effects. Generally, we recommend using smaller designs to save resources, which might also permit more targeted follow-up experiments based on what was learned. When possible, such "sequential" experimentation is preferable to large, one-shot experiments, so conserving resources at earlier stages to allow this is a worthwhile strategy. But there is no guarantee, and if the cost of the additional runs is modest, more runs should be considered.

3.5 Miscellaneous Matters

In this section, we discuss several remaining issues that, while important, don't fit neatly into the previous sections.

3.5.1 Choosing highs and lows

The "high" and "low" settings of the factors in all the designs – which might just label different categories if "high" and "low" are not meaningful – are simply indicated by "+" and "−" in the design tables listed in the appendix. But if the factor is a continuous variable like temperature or concentration, actual high and low values for these settings have to be chosen to run the experiment. How should this be done?

Quite frequently, the choice is dictated by the context. For example, to improve the efficiency of a drying process, the "−" level of the dryer air flow is the current level and the "+" level may simply be as high as it can go. Or maybe a concentration setting for "−" in a chemical process is "the minimum concentration that is theoretically necessary to saturate the solution." These sorts of situations are generally not a problem.

But what if this is not the case? One thing to keep in mind when making such choices is that experimental effects are relative to the ranges over which the variables are changed. After all, if temperature is changed from close to absolute zero to that of the temperature of the sun, it *will* have an effect! And if if it is changed from 100 to 100.000001 degrees, it almost certainly won't. Another point to keep in mind is that in empirical investigations such as these, the results are only trustworthy within the ranges over which the experimentation occurred. Extrapolation beyond is always tempting, but, unlike the situation for more theoretically based work,[6] there is no compelling reason to believe it will be meaningful to do so.

This sets up a certain tension in making the choice. On the one hand, one would like to make the experimental range as large as possible to determine the effect of the change and avoid the need for extrapolation. On the other, the range shouldn't be set so large that it does not realistically reflect practice or go outside the realm of "normal" behavior.[7]

The following comments may offer some guidance. First, in a multifactor experiment, since the purpose of the experiment is to compare the effects of the various factors, it is generally desirable to have the

---

[6]That is, experimentation designed to fix the parameters of a previously determined model; although one might argue that extrapolation beyond the range of the data is still needed to dtermine the validity of the model in the extended range.

[7]e.g. so that "nonlinearity" occurs.

levels set so that, to the best of prior knowledge, one expects changes
in all the factors to have similar effects. Although there is typically
only a rough idea of what such changes should be – otherwise why do
the experiment in the first place? – choosing the values in this way
gives Mother Nature maximum latitude to help sort out the results.

Nevertheless, there are circumstances this would not be appropriate.
For example, in so-called "robustness" experiments, some (even all)
experimental factors are typically varied between values that are not
expected to affect the response specifically to demonstrate that they do
not.[8] Alternatively, lows and highs for a variable may be deliberately
chosen to be somewhat "extreme" to invite problems to occur. This
would be the case, for example, when troubleshooting, where one wants
to deliberately trigger an upset to determine its source. Multifactor
designs are a better way to do this than traditional OFAT studies,
because they allow for the possible interactions that often are the root
causes.

Second, ranges should be chosen large enough so that if a factor does
have an effect it will be seen. Again, this is a chicken and egg problem:
if we already knew how large the effects were, we wouldn't have to do
the experiment! It may be helpful in this regard to remember that
because these designs use "averaging" to assess effects of the changes
(see section 1.2.1 of the Introduction), they are better able to detect
and quantify such effects when they are there. So it is generally not
necessary to make changes as large as intuition might suggest based on
the expected effects of the individual factors . While this is vague, it's
about the best that can be offered. Mostly, good judgment by subject
matter experts is required.

3.5.2   More levels

Two level experiments are useful in a wide variety of experimental
situations because they permit experimenters to evade the Curse of
Dimensionality and economically study many experimental factors simultaneously. This is especially advantageous in complex industrial

---

[8]And that their are no interactions among them that cause unforeseen problems
either.

processes with many possible factors and where unknown interactions among a critical few can result in unexpected behavior.

There are also situations where it may seem that more than two levels are needed, but where two level experiments can be used sequentially to accomplish the same experimental goals more efficiently. Suppose, for example, that it is desired to optimize a process by determining "optimal" settings for what are believed to be the 4 key variables, V1 – V4, that control process results. Four various reasons, it is felt that to do this, V1 and V2 have to be studied at 5 levels each, while V3 and V4 have to be studied at 3 levels each. If all possible combinations of these were included in an experiments, it would require a minimum of $5 \times 5 \times 3 \times 3 = 225$ experimental runs, which is likely ridiculous.

It is also unnecessary: the vast majority of the 225 results would probably be uninteresting. What should almost certainly be done instead is to reduce the experimental range and run an 8 or 16 run design in the 4 factors. This would suffice to identify the important factors and quantify their effects over the reduced ranges. This information would then be used to change the levels to new ones for which improved results would be expected, and then another small experiment at these new levels would be run. This process should then be repeated until no further improvement is possible, at which point the "optimal" region would have been found. Speaking geometrically, the idea is to move the variable settings of the next experiment in the direction of improvement indicated by the previous one. Typically, only two or three such iterations (e.g. two or three 8 run designs) would be required, a vast improvement over the "try everything" approach described above, which would never be implemented anyway, and which would probably have been replaced by an OFAT approach in which the variables were changed and "optimized" one at a time. Unfortunately, one at a time "optimization" can lead to decidedly suboptimal results overall, especially for complex interactive processes.

We will not explore this idea of sequential experimentation further here. Informally, of course, it is just another name for the sort of thing all experimenters do anyway, i.e. the scientific learning process. More formally, it is known as *Response Surface Methodology* in statistics, and readers should consult references (including the Web) for much much more on this topic.

Nevertheless, sometimes more levels are needed and it isn't possible to experiment sequentially. For example, this would be the case if one of the factors in an experiment to improve the "smoothness" of a coating was which of four different additives to use. In this case, "additive" is just a categorical factor with four different levels. Because there is certainly no "high" or "low" among them, the only way to determine how they compare is to experiment with all of them. When there are possibly many such factors and/or when sequential approaches may be a practical impossibility,[9] two level designs don't fit.

More advanced methods sometimes can help finesse these situations; but the simple fact of life is that the Curse of Dimensionality becomes a real obstacle in these sorts of scenarios, making it impossible to sort through all combinations of levels with many factors. The best advice that we can give if you believe you are in this situation is to consult a statistical expert, but sometimes the inherent complexity of the experimental context allows no simple and effective approach.

3.5.3   Center points

This is a concept that only applies when all factors in an experiment are continuous. A definition is:

> **Center Point** An experimental run in which the levels of all experimental factors are set at a value between their "–" and "+" levels.

Unfortunately, this is more of a non-definition – there could be (in-finitely) many such "center points" depending on exactly which "in between" values are chosen. The usual definition fixes the values at exactly the middle of their experimental ranges, thus making the center point unique. But in many circumstances,[10] the *response* would not be expected to be in the "middle" when the experimental factors are. To get a "middle" response, prior knowledge might suggest that

---

[9] ... perhaps due to tight timelines and extended time required to evaluate current results to determine what the next factor level settings should be

[10] Specifically, when the response is a nonlinear function of the variable.

one or more factors might need to be closer to one end or the other. Another way of saying this is that maybe some factors should be in the middle on a logarithmic or exponential scale, not on the original arithmetic scale. This is what the above definition is intended to allow.

The essential reason to include center points can best be understood in geometric terms. A two level design in n continuous variables can be thought of as a choice of n vertices out of the 2^n total that make up a "hypercube" in n dimensions. It's exactly analogous to a 2^3 design in which the 8 vertices are the 8 points with 3-d coordinates (± 1, ± 1, ± 1), except there are n ± 1's, not just 3. Because all the designs in the catalogue have the experimental points at "corners" (e.g. vertices) of the hypercube there is empty volume in the middle where no experimental runs exist – and the bigger n is, the bigger this empty volume. Adding a center point to a design provides a kind of check on the assumption that results in the middle really are in or near the "middle" of the results at the corners.

A natural and important situation in which this kind of experimental check would show results at the center point *not* to be in the middle is when an "optimal" setting for the experimental factors is nearer the middle of the hypercube rather than a corner. In this case, the experimental responses at the corners might turn out to be more or less the same and all worse than at the center. Without the center point, you would never know this. This can also be viewed as showing how center points provide a check on whether the ranges of at least some of the experimental factors have been set too wide – the "+'s" are too high and the "−'s" too low – so that the "interesting" results occur between them in the middle. When this happens, the message is that the highs and lows need to be "compressed" to get the experimental runs within the interesting region where the action is.

However, there's a catch to this. When this sort of behavior is seen, it is natural to ask whether the ranges of *all* the experimental variables need to be compressed, or just *some* of them, which is the usual case. Unfortunately, these designs cannot answer that question[11]. So if subject matter knowledge doesn't fill the gap, all the ranges may have to

---

[11]The technical term is that the quadratic effects of all the factors are *confounded*.

be shrunk and the experiment repeated. There is no such thing as an experimental strategy that is immune to all such possibilities; but a good strategy should incorporate the ability to detect and recover from those most likely to occur as quickly as possible, and that is exactly what center points help to do.

Another important aspect of center points is that they *should* be replicated. There are two reasons for this. First, unlike the vertices, there is no averaging for the center point. They must be explicitly replicated and averaged to reduce the uncertainty due to experimental variability(especially if a result is not as expected). Second, replicated center points provide a kind of experimental quality control. If center point replicates are inconsistent or show some sort of trends over the course of the experiment, this suggests that there may be problems that need to be dealt with to assure that experiment results are reliable and reproducible. To best serve this purpose, the center points should *not* be run in random order when experimental runs are sequential; rather, they should be run at the beginning, end, and evenly in between. This provides a sort of monitoring over the duration of the experiment. Similar ideas should be used in non-sequential experiments, with the details adjusted appropriately.

3.5.4 Create a table of actual experimental settings and check it carefully!

Once a design has been chosen and the high and low values for each factor have been determined, the "+" and "−" symbols in the design tables should be replaced with the actual variable values to produce a complete table of the actual experimental settings. Before doing anything else, check this table carefully to make sure, first, that it is correct(transcription errors can easily occur); and second, that there are no runs that will obviously "fail" – e.g. where a chemical reaction won't occur at all, where a coating won't flow, or where the equipment could be damaged. While it is relatively easy to choose high and low settings to avoid such calamities when only one factor at a time is changed, it is much more difficult when many factors simultaneously vary. This is another example of the importance of choosing good highs and lows – in this case, not too extreme. Multifactor designs offer many advantages, but like any powerful tool, they have to be used wisely.

The design tables in the appendix are given in a convenience order that makes them easy to read, but the actual run order/assignment of runs to experimental units should be randomized as much as possible. Occasionally one may wish to make an exception, however, and one such situation is where a small number – perhaps just one – of the runs are "suspect" when the actual design settings in the table are scrutinized. Obviously, if there is certainty that the runs would fail, the highs and lows need to be reset. However, the suspect runs may fall into a gray area. If they actually are fine, they could even be quite informative, so just defensively reducing experimental ranges may not be the best way to proceed. A fairly sensible alternative then is to run the suspect runs first (in a sequential experiment). If they fail, the experimental highs or lows of offending factors can be changed and the experiment restarted without having wasted unnecessary time and effort, as might have been the case if a randomized run order had been used. If things work, the remaining runs can then proceed, randomized, of course.

Of course this doesn't apply if the experiment isn't sequential. Nevertheless, such "pilot" runs can usually be done in other experimental contexts, too. If worst comes to worst, and it is found that some of the runs produce no information *after* the experiment is completed – for example, a sterile cell culture is found to have been contaminated and ruined – then one may have to analyze the results as they stand, however incomplete. Sometime fancy statistical analysis methods can save the day here, but they can be tricky to apply and even trickier to interpret. And there is no guarantee results can be rescued at all. So whenever possible, it's better to avoid disaster by exercising due diligence at the design stage.

3.6 Chapter Summary

This chapter has set out the principles and methods of a strategy for 2 level multifactor experimentation based on design projectability and

effects sparsity. Effects sparsity – aka The Pareto Principle – posits
that even when many factors are included in a design, typically only
a "vital few" will be sufficiently active to be of interest. The designs
presented here are "equal opportunity"[12] in the sense that all factors
are given the same opportunity to demonstrate their importance; and,
once a subset of critical factors (up to the design projectability) is
identified, restricting the design to just those factors provides not only
all combinations of their highs and lows, but also some inkling of
experimental variability through hidden replication.

So to put this strategy into practice, we need a procedure for determining which are the vital few active factors once the experimental
results are in hand. That is the task of the next chapter.

Despite all this nice machinery, it's important to acknowledge that no
single (or simple) design strategy can fit all experimental needs or be
successful against all possible insults that Mother Nature might visit:
there are certainly circumstances where more sophisticated and complex designs and analyses would do better or may even required. After
all, the experimental design literature is vast exactly for this reason.
We have also deliberately limited the design catalogue to "small" designs (24 runs or less) and a modest number of factors (12 or less). We
have done so because we believe that these suffice for the vast majority
of experiments that are encountered in practice, and we prefer to keep
things as uncluttered as possible. However, both of these limitations
can easily be relaxed within the projectability framework, even using
the designs already provided. The references indicate how.

---

[12]A term due to J. Stuart Hunter

3.7 Appendix: Design Tables

Design Tables

Design tables are given for up to 12 experimental factors in up to 24 runs, listed by number of runs. For k factors, use the first k columns of the design. Note that there are three separate tables for 16 run designs – one for 4 and 5 factors, another for 6 to 8 factors, and a third for 9 to 12 factors.

Note: The runs (rows) in these design tables are given in convenience order. **The order of the rows and/or assignment of experimental units to treatments should be randomized in actual use.**

Table 3.4: 8 Run Designs

	F1	F2	F3	F4	F5	F6	F7
1	-	-	-	-	+	+	+
2	+	-	-	+	-	-	+
3	-	+	-	+	-	+	-
4	+	+	-	-	+	-	-
5	-	-	+	+	+	-	-
6	+	-	+	-	-	+	-
7	-	+	+	-	-	-	+
8	+	+	+	+	+	+	+

Table 3.5: 12 Run PB Designs

	F1	F2	F3	F4	F5	F6	F7	F8	F9	F10	F11
1	+	+	+	+	+	+	+	+	+	+	+
2	-	+	-	+	+	+	-	-	-	+	-
3	-	-	+	-	+	+	+	-	-	-	+
4	+	-	-	+	-	+	+	+	-	-	-
5	-	+	-	-	+	-	+	+	+	-	-
6	-	-	+	-	-	+	-	+	+	+	-
7	-	-	-	+	-	-	+	-	+	+	+
8	+	-	-	-	+	-	-	+	-	+	+
9	+	+	-	-	-	+	-	-	+	-	+
10	+	+	+	-	-	-	+	-	-	+	-
11	-	+	+	+	-	-	-	+	-	-	+
12	+	-	+	+	+	-	-	-	+	-	-

Table 3.6: 4 and 5 Factor 16 Run Designs

	F1	F2	F3	F4	F5
1	-	-	-	-	+
2	+	-	-	-	-
3	-	+	-	-	-
4	+	+	-	-	+
5	-	-	+	-	-
6	+	-	+	-	+
7	-	+	+	-	+
8	+	+	+	-	-
9	-	-	-	+	-
10	+	-	-	+	+
11	-	+	-	+	+
12	+	+	-	+	-
13	-	-	+	+	+
14	+	-	+	+	-
15	-	+	+	+	-
16	+	+	+	+	+

Table 3.7: 6 - 8 Factor 16 Run Designs

	F1	F2	F3	F4	F5	F6	F7	F8
1	-	-	-	-	-	-	-	-
2	+	-	-	-	+	+	+	-
3	-	+	-	-	+	+	-	+
4	+	+	-	-	-	-	+	+
5	-	-	+	-	+	-	+	+
6	+	-	+	-	-	+	-	+
7	-	+	+	-	-	+	+	-
8	+	+	+	-	+	-	-	-
9	-	-	-	+	-	+	+	+
10	+	-	-	+	+	-	-	+
11	-	+	-	+	+	-	+	-
12	+	+	-	+	-	+	-	-
13	-	-	+	+	+	+	-	-
14	+	-	+	+	-	-	+	-
15	-	+	+	+	-	-	-	+
16	+	+	+	+	+	+	+	+

Table 3.8: H2: 9 - 12 Factor 16 Run Designs

	F1	F2	F3	F4	F5	F6	F7	F8	F9	F10	F11	F12
1	-	-	-	-	-	-	-	-	+	-	-	-
2	+	-	-	-	-	+	+	+	-	+	-	-
3	-	+	-	-	+	-	+	+	-	-	+	-
4	+	+	-	-	+	+	-	-	-	-	-	+
5	-	-	+	-	+	+	-	+	+	+	+	-
6	+	-	+	-	+	-	+	-	+	+	-	+
7	-	+	+	-	-	+	+	-	+	-	+	+
8	+	+	+	-	-	-	-	+	-	+	+	+
9	-	-	-	+	+	+	+	-	-	+	+	+
10	+	-	-	+	+	-	-	+	+	-	+	+
11	-	+	-	+	-	+	-	+	+	+	-	+
12	+	+	-	+	-	-	+	-	+	+	+	-
13	-	-	+	+	-	-	+	+	-	-	-	+
14	+	-	+	+	-	+	-	-	-	-	+	-
15	-	+	+	+	+	-	-	-	-	+	-	-
16	+	+	+	+	+	+	+	+	+	-	-	-

Table 3.9: 20 Run PB Designs

	F1	F2	F3	F4	F5	F6	F7	F8	F9	F10	F11	F12
1	+	+	+	+	+	+	+	+	+	+	+	+
2	-	+	-	-	+	+	+	+	-	+	-	+
3	-	-	+	-	-	+	+	+	+	-	+	-
4	+	-	-	+	-	-	+	+	+	+	-	+
5	+	+	-	-	+	-	-	+	+	+	+	-
6	-	+	+	-	-	+	-	-	+	+	+	+
7	-	-	+	+	-	-	+	-	-	+	+	+
8	-	-	-	+	+	-	-	+	-	-	+	+
9	-	-	-	-	+	+	-	-	+	-	-	+
10	+	-	-	-	-	+	+	-	-	+	-	-
11	-	+	-	-	-	-	+	+	-	-	+	-
12	+	-	+	-	-	-	-	+	+	-	-	+
13	-	+	-	+	-	-	-	-	+	+	-	-
14	+	-	+	-	+	-	-	-	-	+	+	-
15	+	+	-	+	-	+	-	-	-	-	+	+
16	+	+	+	-	+	-	+	-	-	-	-	+
17	+	+	+	+	-	+	-	+	-	-	-	-
18	-	+	+	+	+	-	+	-	+	-	-	-
19	-	-	+	+	+	+	-	+	-	+	-	-
20	+	-	-	+	+	+	+	-	+	-	+	-

Table 3.10: 24 Run PBfo Designs

	F1	F2	F3	F4	F5	F6	F7	F8	F9	F10	F11	F12
1	-	-	+	-	-	-	+	+	+	-	+	-
2	+	-	-	+	-	-	-	+	+	+	-	-
3	-	+	-	-	+	-	-	-	+	+	+	-
4	+	-	+	-	-	+	-	-	-	+	+	-
5	+	+	-	+	-	-	+	-	-	-	+	-
6	+	+	+	-	+	-	-	+	-	-	-	-
7	-	+	+	+	-	+	-	-	+	-	-	-
8	-	-	+	+	+	-	+	-	-	+	-	-
9	-	-	-	+	+	+	-	+	-	-	+	-
10	+	-	-	-	+	+	+	-	+	-	-	-
11	-	+	-	-	-	+	+	+	-	+	-	-
12	+	+	+	+	+	+	+	+	+	+	+	-
13	+	+	-	+	+	+	-	-	-	+	-	+
14	-	+	+	-	+	+	+	-	-	-	+	+
15	+	-	+	+	-	+	+	+	-	-	-	+
16	-	+	-	+	+	-	+	+	+	-	-	+
17	-	-	+	-	+	+	-	+	+	+	-	+
18	-	-	-	+	-	+	+	-	+	+	+	+
19	+	-	-	-	+	-	+	+	-	+	+	+
20	+	+	-	-	-	+	-	+	+	-	+	+
21	+	+	+	-	-	-	+	-	+	+	-	+
22	-	+	+	+	-	-	-	+	-	+	+	+
23	+	-	+	+	+	-	-	-	+	-	+	+
24	-	-	-	-	-	-	-	-	-	-	-	+

3.8 References and Resources

The following two references, especially Cheng, provide mathematical derivations of the projection properties of orthogonal arrays that are the basis for the design catalogue. Prior computer exploration that suggested these results was via computer search, and is referenced in these papers.

Box, G.E.P and Tyssedal, J, (1996) Projective Properties of Certain Orthogonal Arrays, *Biometrika*, **83**, 950-955.

Cheng, C-S. (1995) Some Projection Properties of Orthogonal Arrays, *The Annals of Statistics*, **23,** 1223-1233.

The standard "classical" treatment of 2-level screening designs (number of runs a power of 2, reliance on complete confounding and factor aliasing) is well described in:

Box, G.E.P., Hunter, J.S. , and Hunter, W.G. (2005) *Statistics for Experimenters: Design, Innovation, and Discovery*, 2nd Ed., John Wiley and Sons, Hoboken

The definitive reference on Response Surface Methodology written by two of its originators is:

Box, G.E.P., and Draper, N. (2007) *Response Surfaces, Mixtures, and Ridge Analyses*, John Wiley and Sons, Hoboken

This book is large and fairly technical, but the authors offer trenchant commentary throughout on the use and abuse of statistical methods in scientific experimentation that even nontechnical readers will find valuable. If you want the bigger picture of how statistical experimental design works and fits into engineering and technology, this is the book to get.

The projection properties of the Hall 16 run design are derived in:

Box, G.E.P., and Tyssedal, J. (2001) Sixteen run designs of high projectivity for factor screening, *Communi. Statist: Simulation and Computation, 30*, 217-228.

Hall, M.J.(1961) Hadamard matrices of Order 16, *Jet Propulsion Lab. Summary.* *1,* 21-26.

Online versions of all designs except H2 can be found by web search on "Plackett-Burman Designs." Wikipedia™ and the NIST Engineering Statistics Handbook seem to be reasonably convenient sites from which the designs can be extracted.

Chapter 4

Analysis

This chapter gives a step-by-step procedure for analyzing the results of a projectable design to determine the "active" experimental factors – that is, those "vital few" primarily affecting the response(s) – and shows how to visualize and interpret their joint effects. Initially, we'll demonstrate the procedure on a constructed artificial example where we know what the results should be. This helps to provide insight on how things work. But the meat of the chapter will be several examples with real experimental data, which often exhibit subtleties and complexities that challenge straightforward interpretation. This is where it gets interesting.

Before proceeding, we need to note a few points.

- The goal of an analysis is to find the most important factors that affect the response and, if so, to estimate by how much. It's important to keep in mind, however, that just because a factor has been found to be active, it doesn't mean that its impact is large enough to be scientifically or economically meaningful. Only subject matter expertise can provide that judgment. So, for example, one might find that increasing temperature by two degrees does reduce impurities, but the reduction might be too small to be worthwhile. This is particularly the case when, as is typical, there are other responses besides impurities – say energy use, for example – that are a concern.

- Expanding on this last comment a bit, in most industrial processes, there are several, often to some extent competing, responses of interest, with at least part of the reason for experimenting being to find a better trade-off among them. A frequent example in product development is to improve some particular quality attribute while minimizing cost and not adversely impacting others.

The analysis methods presented here are for a single measured response. There are two ways to adapt them to multiple responses. The obvious method is simply to repeat the analysis separately for each response. This is often both appropriate and revealing, because it exploits the ability of projectable designs to study many factors. Some of the factors may turn out to affect one response of interest, while others may affect others. In essence, this provides separate dials to turn to control the different responses. Providing more possible "dials" – experimental factors – to investigate is key to the ability to do this.

However, there are some subject matter specific situations where the different responses are inherently related and need to be combined and analyzed together. Examples include several measurements of drug concentration in blood as part of a "PK" (pharmokinetic) study; heights of peaks of an IR or mass spectrum; or, as an extreme case, the intensity of the separate pixels in an image such as a metallurgical micrograph. Here subject matter expertise typically would provide specific recipes to summarize the response into one, or at most a very few, distinct key characteristics, such as area under a fitted drug concentration curve, average peak height, or average grain size in the micrograph. The analysis would then proceed on these summarized values as before.

There are, however, times when such subject matter driven summarization may not be available or where there might be an interest in supplementing it with alternative "data driven" methods. We wish only to mention here that there are such "multiple response" statistical methodologies available. Unsurprisingly, they are complex, so if you are interested in pursuing such approaches, you'll need to consult a statistical expert.

- Occasionally, the analysis identifies no "vital few" active factors; or, what comes to the same thing, *all* factors appear to be active. There are two reasons why this might happen. First, especially if relatively few (3 or 4, say) variables are included in the experiment and their highs and lows are set too extreme, all may actually have real and roughly comparable effects. Conversely, if the experimental ranges are too narrow, none may, so that the results are basically just experimental noise. With small experiments, it is hard for the experimental data alone to distinguish the "all" from the "none" situations. Sometimes, prior knowledge of what "large" changes in the response should look like can resolve this, but that again requires subject matter expertise external to the experimental results themselves. From a purely statistical perspective, the results are inconclusive.

4.1 The Procedure

We're now ready to give the recommended analysis procedure. We shall first just lay it out and then follow with some point by point comments and a contructed, worked example. The bulk of the chapter will then consist of demonstrating the procedure on several real examples.

1. **Thoroughly examine your results.** Before doing any "statistical" analysis, look over the experimental results carefully. If they are individual measurements, look at them in the design table or in a simple graph. If they are more complex – a curve, a chromatagraph, or an image, for example – lay them out together in an organized way. The goal is simply to see if something is amiss or something doesn't "look right' – a response out of range or way different than anything else (though you need

to be careful here; sometimes such responses are real, revealing unsuspected interactions among the experimental factors), peaks that should not exist, or a blurry or distorted image, for instance. Such aberrations may indicate fundamentally flawed results that should not be included in the analysis and may require re-running.

2. **Examine the center points or other controls if they have been run.** The purpose here is the same as in 1), to see if something is amiss. Are there trends, shifts, unusual values, or other indication of systematic extra-experimental influences? While randomization, normalization, and/or blocking can provide some protection against the corrupting effects of such insults, they cannot guarantee immunity, so knowing about possible problems in order to decide how – or whether – to best proceed is important.

3. **Make main effects plots; calculate and sort the main effects by size.** Use these to select the 3 (for 3-projectable designs) or 4 (for 4-projectable designs) active factors.

4. **Make a "cube" or "trellis" type plot in the active factors.** The specifics will depend on the software that is used. The intent is to visualize how the response changes over the "design space" of the active factors.

5. **Make any needed supplementary plots for presentation.** These include:

    (a) Interaction plots.

    (b) Enhanced main effect plots.

    (c) Cube or trellis-type plots in only the most "important" of the active factors.

    (d) Any other plots that might be informative.

Warning: do *not* put error bars or any other indication of statistical uncertainty on any of the plots.

4.2 Demonstration on a constructed example

To show how this all fits together, we apply the procedure on a constructed example in which the "truth" is known. Our made-up example is a 12 run, 6 factor 3-projectable design with response, y, "Truth," given by the following equation:

$$y = 92 + 3 \times F2 - 2 \times F5 + F2 \times F5$$

This says that only 2 of the 6 factors, F2 and F5, are active. The $F2 \times F5$ term indicates that there is a modest interaction between the two factors. For the 4 possible ±1 combinations of the factors, the expected values of y from this formula are:

F2 = -1; F5 = -1: $y = 92$ [92 + 3×(-1) − 2×(-1) + (-1)×(-1)]

F2 = +1; F5 = -1: $y = 96$ [92 + 3×(+1) − 2×(+1) + (+1)×(-1)]

F2 = -1; F5 = +1: $y = 86$

F2 = +1; F5 = +1: $y = 94$

(The calculation details are shown in brackets only for the first two values).

For the center point, F2 = 0 and F5 = 0, the expected value is 92, of course.

Some random error has been added to these "true values" to simulate experimental and measurement variability, resulting in the following table of experimental "results" that is to be analyzed:

F1	F2	F3	F4	F5	F6	y
−	+	−	+	+	+	96.2
+	−	+	+	+	−	85.8
−	+	+	+	−	−	95.8
+	+	+	−	−	−	93.5
+	+	−	−	−	+	95.4
+	−	−	−	+	−	85.9
−	−	−	+	−	−	93.1
−	−	+	−	−	+	91.4
−	+	−	−	+	−	93.2
+	−	−	+	−	+	90.4
−	−	+	−	+	+	86.3
+	+	+	+	+	+	95.6
0	0	0	0	0	0	91.4
0	0	0	0	0	0	92.5
0	0	0	0	0	0	93.4

Note that the table has been displayed in a convenience order for clarity, but should actually have been run with corner points in random order and the center points at the beginning, middle, and end, as previously discussed.

Now we play the game by seeing how well the analysis procedure does in "recovering" the truth given by the equation, which, in reality, is always unknown (and for which simple equations like this would just be an approximation, anyway).

1. & 2. Since this is just a made-up example, there's no real subject matter knowledge to apply, and there's nothing that appears noteworthy in the data.

3. Main effects plots for the 6 factors are shown in Figure .4.1

For each of the 6 factors, this is simply a plot of the y responses versus the 3 settings of the factor. Lines are drawn connecting the average y value at each setting. "Active" factors are those which show the largest change between the average − and + settings of the factor. F2 clearly shows the largest change, but it is not clear what others should be included. Note also that the center point values here are roughly in the middle where they "should" be, showing no indication of curvature.

Figure 4.1: **12 Run, 6 Factor Example: Main Effects Plots**

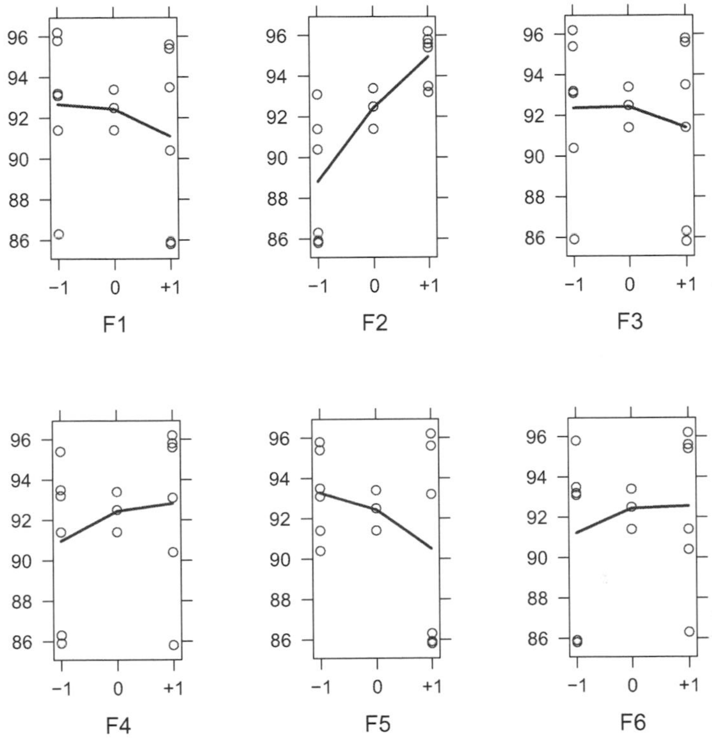

Main Effects: Main effects are calculated as the average of the (6) results at the + setting minus the average of the (6) results at the − setting for each of the factors. In other words, these are the values that are graphed in the main effects plots. They are (the details are shown for the first two):

F1: -1.55 [91.1 − 92.65]
F2: +6.1 [94.9 − 88.8]
F3: -.95
F4: +1.9

F5: -2.8
F6: +1.35

So the three largest magnitude (ignoring sign) effects are F2, F5, and F4 in decreasing size.

4. Plot the results in the three active factors, F2, F4, and F5.

We do this in two slightly different ways, first using a traditional "cube" plot; and second using a slightly more graphical "trellis-with-glyphs" plot. At present, only cube plots are widely available in software; however, the trellis plot is not difficult to construct manually, especially if the glyphs (the filled bars) are replaced by text giving the value. Use whichever you prefer.

The cube plot represents the triplets of factor settings for the three

Figure 4.2: Cube plot of F2, F4, and F5

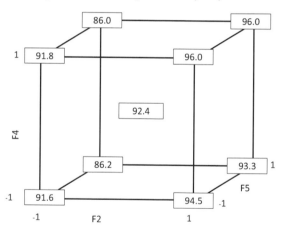

factors as points on a cube, with the value shown at each point being the average y value at that settings. So, for example, the left, upper, front face value of 91.8 is the average of the two y values of 93.1 and 90.4 when F2 $= -$, F4 $= +$, and F5 $= -$, just slightly off due to rounding. The 92.4 hanging in the center is the average of the 3 center point values, of course. Note that for half the corner points, the "average" is of just one value.

Study of this plot first clearly shows the F2 effect: all the results on the right side are higher than on the left. On the right face, it appears that higher value y values are associated with the + setting of F4. Of course, we "know" that this is really just experimental noise. Less clear, perhaps is the interactive effect: when F2 is at its "−" setting on the left side of the cube, increasing F5 is results in a large decrease in y. When F2 is + on the right side, there is hardly any change when F5 is increased. If the goal was to maximize y, clearly F2 should be + and possibly F4 should also be at its + setting.

Now let's look at an alternative way to view these results via a trellis plot.

Figure 4.3: Trellis plot of F4, F2, and F5

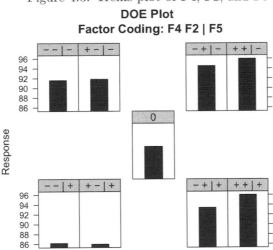

The trellis plot needs some explanation. First, the nine design settings are now laid out "informatively" in separate "panels." The setting for each panel is coded by the gray key at the top of the panel. The plot heading gives the factors ordering, in this case F4, F2, F5. This means, for example, that the lower left panel with key $- - I +$ shows the predicted response when F4 and F2 are $-$ and F5 is $+$. The vertical bar shows that the first two factors vary horizontally, while the third varies vertically, i.e. F5 is constant in each row. Although the actual settings are obvious in this constructed example, in general, they should be decoded in a separate table giving the true factor levels corresponding to $-$ and $+$.

The panels are ordered so that the leftmost factor varies the fastest and the rightmost the slowest as one moves from left to right and top to bottom. The plots are spaced so that within each dimension, the spacing increases for each group of factor settings. So in this simple 3 factor plot, $-$ and $+$ settings for F4 are next to one another horizontally with F2 and F5 constant in each pair; then the pairs are horizontally separated by additional spacing (plus the spacing for a center setting if there is one) corresponding to the F2 setting. Similarly, the two rows are for the $-$ and $+$ settings of F5 with panels in the same column having the same settings of F2 and F4. This arrangement was chosen so that the factor associated with the smallest changes is shown with its changes as close together visually as possible. This facilitates visual assessment – smaller changes are more easily compared when they are closer together.

While for this simple example it is fairly easy to see what is going on with pretty nearly any arrangement, the advantage of this approach is that it extends to more factors (5 or 6 can still be effectively visualized, especially when the viewer has the ability to vary the plot arrangement interactively) and more complex responses. For example, instead of a bar representing a single number, the response might be a curve (e.g. a spectrogram or dose-response curve).[1]

The F2 effect is clearly shown in the horizontal panel pairs. The F2 $\times$ F5 interaction is also clear as the larger increase of bar height when F2 goes from $-$ to $+$ in the top panel pairs (F5 $= +$) than in the bottom (F5 $= -$). As before, it is clear that maximal *predicted* y response requires F2 $= +$, and that once this is done, F2 and F4 have only slight effects, at best.

---

[1]DOEdisplay, an R software package that implements these ideas, will soon be – or already is – available.

5. Another way of plotting the results is to plot the main effects for F4 and F5 again, but this time using plotting symbols distinguished by the level of F2, the largest main effect. Color generally works better for making the distinction, but of course can't be used here in a black and white format. Also, we omit the center points from the plot, as they do not contribute any information in this way of graphing the data.

Figure 4.4: Interaction Plots for F2

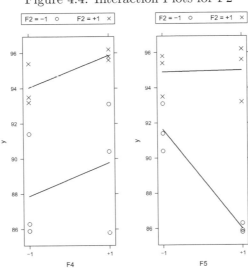

The lines in the plots of Figure 4.4 now connect the average of the (3) y values grouped by F2 level at each level of the factor being plotted. As indicated in the title, this is known as an *interaction plot* because it is a visualization of the (2-way) interactions of F4, and F5 with F2: the effect of F4 is essentially the same at both levels of F2; but the F5 effect depends critically on the level of F2 – it is nil for F2 = + and large and negative for F2 = −.

Of course, one could make similar plots with F5, say, in place of F2. Examination of the main effects plots for grouping in the points often suggests which factors are may have important interactions, but, as here, those factors with dominant main effects are typically those which tend to interact. Careful examination of the cube or glyph plots will always give this information, but if in doubt, it is always possible to make all possible interaction plots to clearly visualize the situation. There are only three for 3-projectable designs (and six for 4-projectable designs).

Summarizing all this, the analysis indicated that F2 is the dominant factor, and there also appears to be a large F2 × F5 interaction. Additionally, F4 may have a modest (additive) effect, at most. Of course, we know that this is correct for F2 and F5, but that there is no F4 effect – it is just "noise." This leads to an important and ultimately difficult question: for a k-projectable design, should all the most important k factors be kept in the final analysis, or should some be eliminated if they do not seem to be "real"? So in this example, should we stop at just including F2 as the only active factor, F2 and F5 only, or are all three needed?

This is actually an example of what is known as the "bias-variance tradeoff" in statistics. In this case, it manifests itself as: how can one determine the proper balance between incorrectly omitting important variables, which leads to "somewhat" inaccurate predictions, versus incorrectly including variables that have no real effect, which lead to excess "noise" in predictions. There's no simple answer to this question, and in fact a large part of (often complex) statistics is devoted to it. For our purposes here, we rely on subject matter knowledge and the gold standard of science: follow-up studies to confirm and extend the results.

In any case, this analysis did correctly extract the essential features of the process that generated the data, leaving just a few details that might require further resolution. More sophisticated statistical analyses[2] could provide additional insight if appropriately applied. But the message we wish to convey here is that these simple graphical procedures usually suffice in practice, especially when subject matter knowledge, always a key part of the mix, is brought to bear.

The real examples that follow will further illustrate these approaches and also show how they work with 4-projectable designs (where, as

---

[2]In particular, multiple regression analysis

one would expect, the complexity increases). Inevitably, we focus here on the *quantitative* aspects of the analyses: what are the sizes of the estimated effects? However, in practice, often the focus is qualitative: *which* of the effects are the vital few? – are they positive or negative? – are their effects cumulative or is their an indication of synergy or interference (interactions)? Once these questions are answered, prior theory and experience then kick in to provide detailed quantitative understanding beyond that contained in the experimental data. Obviously, this is not something that can be shown in the examples, but it is something to keep in mind. Indeed, it is our hope that the graphical analyses promote such further insight, as this is often the most important outcome of the experiment.

4.3 A Cell Culture Experiment

So-called Biological Drugs – "Biologics" for short – are drugs made by inserting engineered genes into cells and then growing the cells in large scale cell culture. When this is done in the "right" way, the growing cells will express the protein products for which the inserted genes code[3], which are then purified into drugs. Grape fermentation, in which the cells are yeast and the "druglike" product of interest is alcohol, is an ancient and somewhat similar idea, though nowhere near as complex, of course.

Engineers developing an improved version of such a process wanted to investigate the effects of six controllable factors on process performance. They were Temperature, pH, atmospheric Pressure in the fermentation vessel, Batch Volume, Volume of the Initial Cell Inoculate, and Ramp (the rate at which certain "ingredients" were added to the culture). Titer and the percentage of certain impurities were the measured responses of interest, but we shall focus in this example on only one, which we label simply as "Impurity1" to avoid any proprietary issues.

Although the experiment was to be run only at small scale in parallel small fermenters, the number of fermenters available was limited,

---

[3]Often antibodies

the cell culturing process required approximately 2 weeks, and then the fermentation results has to be purified and measured, which required several weeks further, depending on the laboratory availability. Hence, it would take almost two months at minimum to complete and get back the results from even a modest experiment. Based on these conisderations, it was decided to run a 3-projectable,12 run PB design with 2 added center points. Table 4.1 gives the design and Impurity1 results. However, again to avoid proprietary issues, the actual settings of all factors and the response have been scaled, and units of measurement are omitted. This just hides the true settings and magnitudes of the effects, but the relative effect sizes are unchanged, and that is what is of interest here.

Table 4.1: 6 Factor Fermentation PB12

	Temp	pH	Pressure	BatchVol	InocVol	Ramp	Impurity1
1	38	7.55	0.05	13.6	60	10	3.86
2	38	7.45	0.35	13.8	100	-10	6.14
3	36	7.55	0.05	13.8	100	10	7.20
4	38	7.45	0.05	13.6	100	-10	6.72
5	36	7.55	0.35	13.8	60	-10	4.08
6	36	7.45	0.35	13.6	60	10	3.94
7	38	7.55	0.35	13.6	60	-10	4.66
8	36	7.45	0.05	13.8	60	-10	7.73
9	36	7.45	0.35	13.6	100	10	6.93
10	36	7.55	0.05	13.6	100	-10	6.28
11	38	7.45	0.05	13.8	60	10	7.16
12	38	7.55	0.35	13.8	100	10	4.54
13	37	7.5	0.2	13.7	80	0	5.59
14	37	7.5	0.2	13.7	80	0	6.95

The main effect plots are given in 4.5. Note the relatively large variability in the two replicate center points, which indicates that we should avoid over-interpreting the results. The plots show that pH and Pressure have the largest in magnitude effects. It looks like InocVol is third. This is confirmed by calculating the six effects as the difference in average Impurity between high and low levels of the factors giving: Temp = -.51, pH = -1.33, Pressure = -1.44, Batch Volume = .74, Inoculation Volume = 1.06, and Ramp = -.33 . So Pressure, pH, and Inoculation Volume are, indeed, the three largest magnitude effects.

Figure 4.5: Main Effect Plots for Plackett-Burman Cell Culture Design

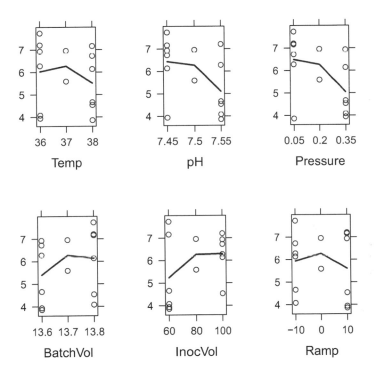

The design is 3-projectable, so a trellis plot in pH, Pressure, and Inoculation Volume is given in Figure 4.6. Given that each of the results shown is an "average" of either one or two values and that the difference between the two replicate center values was 1.5, the impurities can be characterized as being essentially either large or small. There were four setting that were small and five, including the center point, that were large.

Figure 4.6: Trellis Plot of %Impurity for pH, Pressure, and InocVol

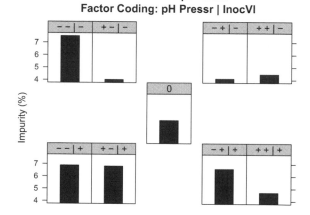

It is evident that three of the four runs where InocVl = Innoculation Volume was at its − setting of 60 had the lowest impurity levels. Unfortunately, the fourth, where pH and Pressure were also low, had the highest. So a reasonable summary is: to minimize the impurity, use low innoculation volume so long as pH and pressure are not both near their low values. If higher innoculation volume must be used, then pH and pressure should be kept as high as possible (the +++ result at bottom right).

4.4   A 4-Projectable Design for a Wave Bioreactor

A wave bioreactor is a liquid media filled bag that is used to culture cells that have been engineered to secrete biotherapeutic products.

The bag is attached to an oscillating rocker that agitates the liquid and produces "waves" in it, thus assuring the even distribution of oxygen and nutrients to the cells.

It was desired to determine the conditions that would maximize the dissolved oxygen content of the media. Five factors were investigated: media volume in bag as a percentage of bag volume; air flow rate (l/minute); maximum rocking angle (in degrees); rocking rate (rocks/minute); and air temperature ($^{\circ}$C). A 4-projectable, 16 run design with six center points was run. The design and results are shown in Table 4.2.

A quick scan of the results shows something interesting: almost all dissolved oxygen values were below about 30, except for one at 63 and two others at about 110. This suggests that something interesting is going on, though it may not be clear exactly what at first glance. Let's proceed with the standard analysis to see.

Main effects plots are shown in Figure 4.7. Clearly, volume is the major actor, with higher oxygen strongly associated with lower volume. For the other four factors, the trend isn't as clear. However, note that for rock angle and rate, the two high oxygen results occurred at one setting of each only – the high rock angle and rate; while for flow rate and temperature, the high oxygen results were split between the two settings. Although not conclusive, when unusual results are associated with specific settings of some of the factors, it suggests that there may be synergy – that is, an interactive effect – occurring among the factors. We can investigate this with the trellis glyph plots. The main effects were: Volume = -45, Flow Rate = .8, Rock Angle = 21.1, Rock Rate = 33.4, and Temperature = 5.9. Because the design is 4-projectable, we only need to exclude Flow Rate from the trellis plot, which is shown in Figure 4.8.

Table 4.2: Wave Bioreactor Design and Results

Volume	Flow Rate	Rock Angle	Rock Rate	Temp- erature	Dissolved Oxygen
30	.3	7	22.5	37	27.8
30	.3	7	22.5	37	27.5
50	.3	4	30	35	11.0
10	.3	10	30	35	108.9
10	.3	10	15	39	38.6
50	.1	4	15	35	4.5
10	.1	4	30	35	48.0
10	.1	4	15	39	32.7
10	.1	10	30	39	117.3
50	.5	4	15	39	3.7
30	.3	7	22.5	37	27.4
30	.3	7	22.5	37	28.7
30	.3	7	22.5	37	23.9
50	.1	10	15	39	7.9
10	.1	10	15	35	31.4
50	.1	4	30	39	12.8
50	.5	10	15	35	8.0
50	.1	10	30	35	24.9
10	.5	4	15	35	22.3
10	.5	4	30	39	63.4
50	.5	10	30	39	30.1
30	.3	7	22.5	37	21.7

Figure 4.7: Wave Bioreactor Main Effects Plots

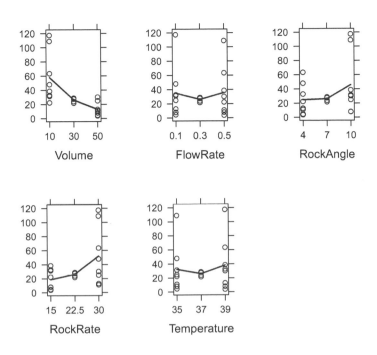

Figure 4.8: Trellis Plot for Wave Bioreactor Experiment

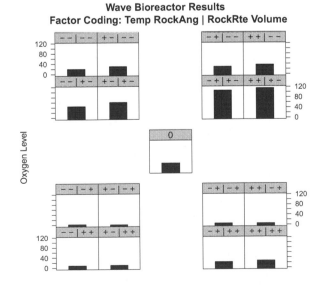

The plots have been ordered so that only temperature changes in each horizontal pair. Clearly the temperature effect is negligible throughout. On the other hand, the remaining 3 factors do show some differences: in horizontal pairs for rock angle; vertically between the rows for rock rate; and top to bottom for volume. Even though it is fairly evident what is going on from th display, we redo the plot in just these three active factors to add clarity, as in Figure 4.9.

Figure 4.9: Trellis Plot in Three Active Factors

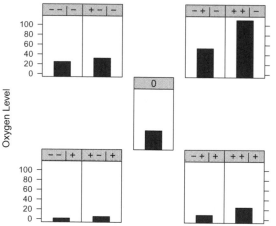

After the fact, the explanation for these results seems straightforward: the higher rock rate and angle represents a more vigorous mixing of the air and fluid in the bag, so more oxygen gets mixed in. With a lesser volume of liquid, there is also more room in the bag for the liquid to get shaken and mixed with the air, or, to put it another way, the air to liquid ratio is higher. It's exactly like shaking up a container of unhomogenized orange juice – the orange solids and liquid mix better when the container is partly empty rather than full. Except in this case it's the air mixing with the liquid, not the solids.

What is interesting is that before doing the experiment the experimenter actually thought that the air flow rate would be the dominant factor, and it it is certainly true that if it were 0, there would be no oxygen at all. But this experiment showed that within the range of air flows used, even the minimum flow rate was sufficient to provide all the oxygen that the cells needed. Once that occurred, what mattered was how well the air mixed, which was determined by fluid volume, RPM, and rocking angle.

4.5 A 15 Factor, 20 Run Design

Soliman *et. al.* (see References*)* describe a 15 factor experiment to optimize the production of polyglutamic acid (PGA) by a *Bacillus* in a bioprocess. Although we have shown only up to 12 factors in the PB20 designs in the design catalogue, Table 3.3, in fact PB20 designs can be used to experiment with up to 19 factors. In this experiment, the authors used a PB20 for 15 factors –13 ingredients of the bacillus growth media and 2 process factors (pH and agitation). The designs and PGA responses are shown in Table 4.3. The identies and low and high settings of the 15 factors are given in the reference.

110

CHAPTER 4. ANALYSIS

Table 4.3: 15 Factor 20 Run PGA Example

Run	X1	X2	X3	X4	X5	X6	X7	X8	X9
1	−	−	−	+	−	+	−	+	+
2	−	−	+	+	−	+	+	−	−
3	+	−	+	+	−	−	−	−	+
4	−	+	+	+	+	−	−	+	+
5	+	+	−	−	+	+	−	+	+
6	−	−	+	−	+	−	+	+	+
7	−	−	−	−	−	−	−	−	−
8	−	+	−	+	−	+	+	+	+
9	−	+	+	−	+	+	−	−	−
10	−	+	−	+	+	+	+	−	−
11	+	+	+	−	−	+	+	−	+
12	−	−	−	+	+	−	+	−	+
13	−	+	+	−	−	−	−	+	−
14	+	+	+	+	−	−	+	+	−
15	+	+	−	+	+	−	−	−	−
16	+	−	−	+	+	−	+	+	−
17	+	+	−	−	−	−	+	−	+
18	+	−	+	+	+	+	−	−	+
19	+	−	+	−	+	+	+	+	−
20	+	−	−	−	−	+	−	+	−

Run	X10	X11	X12	X13	X14	X15	PGA
1	+	+	−	−	+	+	5.16
2	−	−	+	−	+	−	6.88
3	−	+	−	+	+	+	2.70
4	−	+	+	−	−	−	17.86
5	−	−	−	−	+	−	22.78
6	+	−	−	+	+	−	17.22
7	−	−	−	−	−	−	1.16
8	−	−	+	+	−	+	33.54
9	−	+	−	+	−	+	8.28
10	+	+	−	+	+	−	21.56
11	+	−	−	−	−	+	20.30
12	+	+	+	−	−	+	18.36
13	+	−	+	+	+	+	18.96
14	+	+	−	−	−	−	32.68
15	+	−	+	−	+	+	11.08
16	−	−	−	+	−	+	31.96
17	−	+	+	+	+	−	23.72
18	+	−	+	+	−	−	2.84
19	−	+	+	−	+	+	27.26
20	+	+	+	+	−	−	16.52

The analysis proceeds as usual. First, the main effects plots are shown in Figure 4.10.

Figure 4.10: PGA Main Effects Plots

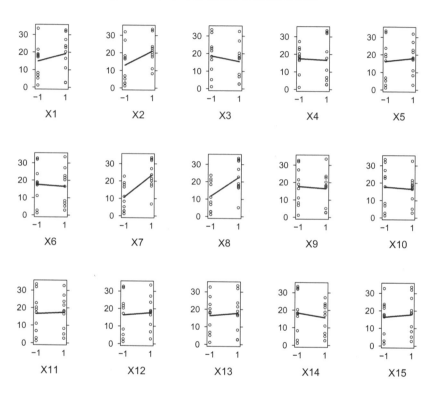

The PB20 is 3-projectable, and X2, X7, and X8 appear to be the largest three effects. With this many effects, rather than listing them all to compare them, it is simpler and more useful to view them graphically with a *Pareto Plot*. This is simply a bar plot of the absolute values of the effects in order from largest to smallest, as shown in Figure 4.11. It is evident that the main effects of X2, X7, and X8 are, indeed largest and that, moreover, the remaining 12 effects are considerably smaller and largely indistinguishable from noise.

Figure 4.11: Pareto Plot of Main Effects

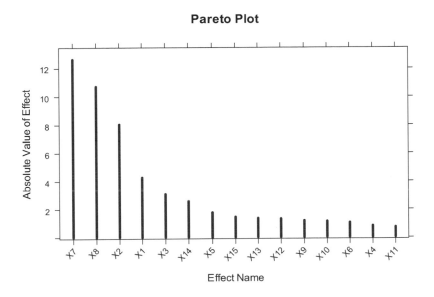

A trellis plot of the PGA results in these three factors is shown in Figure 4.12. We have chosen to show the 20 actual data points, not just the predicted values, to reemphasize that with good experimental design, analysis is typically straightforward. In this case, increases in factors 2,7, and 8 resulted in increases in PGA that were similar for all 3 factors. The plot also shows no evidence of interaction: that is, the effects are additive.

Figure 4.12: Trellis Glyph Plot of PGA Results in X2, X7, and X8

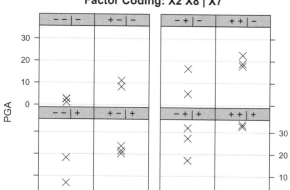

In addition to these three factors, the authors concluded (using multiple regression analysis) that X1 should be included. However, for technical reasons[4], their basis for doing so was flawed, so that X1 may not be as important as claimed. Even if they are correct, the X1 effect is small at best, and the simple projectability strategy demonstrated here identified all factors that the authors found and judged likely to be scientifically the most important.

4.6   Chapter Summary

This chapter has proposed a simple, five step procedure for the analysis of results of 2-level projectable designs. Aside from some arithmetic, the strategy is entirely graphical, requiring no elaborate statistical software or analysis. The key to this simplicity is the projectability of the designs, which, due to the Pareto Principle, allows for straightforward identification and visualization of the "vital few" important factors that are likely to have the greatest effect on the response.

---

[4]Individual t-tests at P = .05 were used for testing the "significance" of factors without adjusting for multiplicity.

Of course, there is no guarantee, and there are circumstances where the strategy can come up short or even produce spurious results – when there are "too many" (more than the design projectabiity) important factors or some interactions are larger than individual main effects, for example. But all strategies for studying many factors in few runs are susceptible to these problems as the inevitable consequence of the extreme economy employed: there is simply too little information to detect such anomalies when few runs are used to study many factors. Of course, classical OFAT design strategies, which tacitly assume away all interactions, are even more vulnerable to such ills, and in any case are too inefficient to even allow the possibility of determining the effects of many factors with reasonable experimental effort.

Fortunately, over 90 years of experience have shown that such problems are rare. Especially when informed by subject matter insight, multifactor experimental design has proven to be an effective strategy in industry, agriculture, systems modeling and simulation, and other areas where complex processes involving many factors need to be studied and improved. We hope that the simple framework for analysis presented here will make this strategy more accessible and encourage wider application. DOE should not just be a tool for experts.

4.7 References and Resources

Soliman, N.A., Berekaa, M.M., and Abdel-Fattah, Y.R. (2005). "Polyglutamic acid (PGA) production by Bacillus sp. SAB-26: application of Plackett–Burman experimental design to evaluate culture requirements." *Applied Microbiology and Biotechnology.* 69: 259-267.

Index

Made in the USA
Middletown, DE
24 August 2019